# StataQuest 4

**Statistics ■ Graphics ■ Data Management**

*Software developed by the Stata Corporation*

*Text by* **J. Theodore Anagnoson**
*California State University, Los Angeles*

**Richard E. DeLeon**
*San Francisco State University*

🔺 **Duxbury Press**
An Imprint of Brooks/Cole Publishing Company
I(T)P® An International Thomson Publishing Company

Belmont, CA • Albany, NY • Bonn • Boston • Cincinnati • Detroit • Johannesburg • London • Madrid •
Melbourne • Mexico City • New York • Paris • Singapore • Tokyo • Toronto • Washington

*Editor:* Curt Hinrichs
*Assistant Editor:* Cynthia Mazow
*Editorial Assistant:* Martha O'Connor
*Production:* Robin Gold/Forbes Mill Press
*Print Buyer:* Stacey Weinberger
*Permissions Editor:* Peggy Meehan
*Copy Editor:* Robin Gold
*Cover Designer:* Stuart Paterson/Image House
*Printer:* Malloy Lithographing Inc.

Printed in the United States of America
1 2 3 4 5 6 7 8 9 10

For more information, contact Duxbury Press at Brooks/Cole Publishing Company, 511 Forest Lodge
Road, Pacific Grove, CA 93950, or electronically at http://www.thomson.com/duxbury.html

International Thomson Publishing Europe
Berkshire House 168-173
High Holborn
London, WC1V 7AA, England

International Thomson Editores
Campos Eliseos 385, Piso 7
Col. Polanco
11560 México D.F. México

Thomas Nelson Australia
102 Dodds Street
South Melbourne 3205
Victoria, Australia

International Thomson Publishing Asia
221 Henderson Road
#05-10 Henderson Building
Singapore 0315

Nelson Canada
1120 Birchmount Road
Scarborough, Ontario
Canada M1K 5G4

International Thomson Publishing Japan
Hirakawacho Kyowa Building, 3F
2-2-1 Hirakawacho
Chiyoda-ku, Tokyo 102, Japan

International Thomson Publishing GmbH
Königswinterer Strasse 418
53227 Bonn, Germany

International Thomson Publishing Southern Africa
Building 18, Constantia Park
240 Old Pretoria Road
Halfway House, 1685 South Africa

ISBN 0-534-52137-1     (Windows 95 version)
ISBN 0-534-26546-4     (Windows version)
ISBN 0-534-26545-6     (DOS version)

StataQuest is a subset of Stata®. Stata is available for a variety of computers, including Windows, DOS,
Macintosh, and UNIX systems. Stata is a registered trademark of:
  Stata Corporation
  702 University Drive East
  College Station, TX 77840

# Contents

■ **Getting Started**

**1**    **Getting Started**                                                                          **1**
What is StataQuest  1
Requirements and Limits  1
Installing, Starting, and Stopping StataQuest  3
Configuring StataQuest to Your Preferences  5
Basic Rules—Windows  9
Basic Rules—DOS  13
Getting Help  13
How to Place Results Window Text in Your Word Processor  14
How to Print Your Work  14
Printing StataQuest Graphs  15
Telling the Difference Among StataQuest's Files  16
StataQuest's Utilities  17
Structure of a Typical StataQuest Session  18
What is Really Happening When You Choose Commands From
   StataQuest's Menus?  18
What's Next?  19
Exercises  19

■ **Data**

**2**    **Entering Data into StataQuest**                                                       **20**
The StataQuest Editor  20
Inputting Your Own Data Through the StataQuest Editor  21
Inputting Your Own Data Through External (Text) Files  27
Creating Random Variables  28
Stata Files  30
Spreadsheet and Statistical Package Files  31
Labeling Your Data  31
Renaming Variables  33
Sorting  33
Exiting the Editor  34
Exercises  34

**3**    **Modifying Your Data**                                    **36**
Why Change Data? 36
Changing Cell Contents 37
Adding Variables or Observations 40
Deleting Variables or Observations 42
Formulas and Functions 43
Recoding Data 47
Subsetting Data 50
Exercises 53

**4**    **Summarizing and Examining Your Data**              **54**
The Basics of Data Analysis 54
What Is in a Data Set? 54
Basic Statistics 60
Exercises 67

**5**    **Tables: Frequency Distributions and**
         **Cross-Tabulations**                                      **69**
What are Frequencies and Cross Tabulations? 69
Frequency Distributions 70
Cross-Tabulation 72
Cross-Tabulation by the Categories of a Third Variable 76
Exercises 78

■ **Graphs**

**6**    **Histograms and Normal Quantile Plots**             **80**
Graphs for Studying the Shapes of Distributions 80
Histograms in DOS versus Windows 81
Continuous Variable Histograms 81
Discrete Variable Histograms 88
Continuous Variable Histograms by Group 89
Discrete Variable Histograms by Group 91
Normal Quantile Plots 92
Exercises 94

**7**    **Stem-and-Leaf Plots and Dot Plots**                **96**
Ways to Graph and See Your Data 96
Stem-and-Leaf Plots 97
Dot Plots 105
Saving and Printing 108
Exercises 110

**8**    **Box Plots and Box and One-Way Plots**             **111**
Graphs for Seeing Both Forest and Trees 111
Box Plots 112
Box Plot and One-Way Plot 114
Box Plot by Group 115
Box Plot Comparisons of Different Variables 117
Box Plot and One-Way Plot Comparisons 119
Exercises 122

**9**     **Bar Charts**                                                                **123**
How Much? Graphs for Comparing Variables and Groups  123
Bar Charts: Totals and Means  124
Bar Charts by Group  124
Bar Chart Comparisons  126
Bar Chart Comparisons by Group  129
Exercises  133

**10**    **Scatter Plots**                                                             **134**
What are Scatter Plots?  134
Plot Y vs. X  137
Plot Y vs. X, Naming Points  138
Plot Y vs. X, with Regression Line  140
Plot Y vs. X, Scale Symbols to Z  142
Plot Y vs. X, by Group  143
Scatter Plot Matrix  145
Exercises  148

**11**    **Time Series Plots**                                                         **150**
Graphing Trends and Change  150
Plot Y vs. X  151
Plot More than One Y vs. X (without Scaling Option)  153
Plot More than One Y vs. X (with Scaling Option)  154
Plot Y vs. Obs. No.  155
Plot More than One Y vs. Obs. No.  157
Exercises  159

**12**    **Quality Control Charts**                                                    **161**
What Are Quality Control Charts?  161
Control (C) Chart for Defects  162
Fraction Defective (P) Charts  164
X-Bar Charts  164
Range (R) Charts  167
X-Bar + R Charts  168
Exercise  171

■ **Statistical Tests**

**13**    **Parametric Tests**                                                          **172**
What are Parametric Tests?  172
One-Sample t-Test  172
Two-Sample t-Test  178
Paired t-Test  180
One-Sample and Two-Sample Tests of Variance  180
Testing for Normality  182
One-Sample Test of Proportion  184
Two-Sample Test of Proportions  187
Exercises  188

**14    Nonparametric Tests**                                               **190**
What Are Nonparametric Tests?  190
Sign Test  190
Wilcoxon Signed-Ranks Test  194
Mann-Whitney Test  196
Kruskal-Wallis Equality of Populations Test  197
Kolmogorov-Smirnov Tests  199
Summary  201
Exercises  202

**15    The Statistical Calculator**                                        **204**
What's a Statistical Calculator?  204
One-Sample Normal Test  204
Two-Sample Normal Test  206
One-Sample t-Test  207
Two-Sample t-Test  208
One-Sample Test of Proportion  210
Two-Sample Test of Proportion  213
One-Sample and Two-Sample Test of Variance  215
Confidence Interval for the Mean  215
Binomial Confidence Interval  217
Poisson Confidence Interval  219
Statistical Tables (Normal, Student's-t, F, Chi-Squared,
    Binomial, Poisson)  219
Inverse Statistical Tables (Normal, Student's-t, F, Chi-Squared)
    (Windows Only)  221
Standard Calculator  222
RPN Calculator (Windows Only)  223
Checking a Cross-Tabulation Printed in a Book  224
Exercises  224

■ **Correlation, Regression & ANOVA**

**16    Correlation**                                                       **226**
The Meaning of Correlation  226
Pearson Correlation Coefficient  227
Spearman Correlation Coefficient  230
Exercises  233

**17    Simple Regression**                                                 **235**
Modeling Two-Variable Linear Relationships  235
Simple Regression  236
Post-Regression Diagnostics  241
Exercises  250

**18    Multiple Regression**                                               **252**
Ordinary Regression with Two or More X Variables  252
Multiple Regression and Regression Diagnostics  253
Stepwise Regression with Forward or Backward Selection  257
Using the *fit* Command to Obtain AV Plots and Influence Statistics  260

Robust Regression  264
Exercises  266

**19**   **Logistic Regression (Windows Only)**          268
What Is Logistic Regression?  268
The Logistic Regression Model and Assumptions  269
Using SQ's Logistic Regression: An Example  272
Graphing the Results  277
Exercises  279

**20**   **Analysis of Variance (ANOVA)**                280
StataQuest's ANOVA Command  280
Nonparametric Analysis of Variance  280
One-Way ANOVA  282
Two-Way ANOVA  285
Repeated Measures ANOVA  287
N-Factor ANOVA & ANOCOVA  290
Exercises  292

**Appendix A: StataQuest's Functions**                 293

**Bibliography**                                        297

**Index**                                               300

■   **List of Tips**
   **1**   Data in Memory versus Data on Disk  22
   **2**   Valid Variable Names  33
   **3**   How to Use Command Mode  41
   **4**   When Is a Number NOT a Number?  48
   **5**   Alphanumeric (String) Variables in Statistical Routines  59
   **6**   Row vs. Column Percentages in Cross-Tabulations  74
   **7**   Handling Overflows in Stem-and-Leaf Plots  105
   **8**   Using String Variables to Identify Outliers in Box Plots  116
   **9**   Use Jittering to See More Detail in One-Way Plots  116
   **10**  "Stacked" Bar Charts  131
   **11**  Creating Bar Charts for One or a Few Observations  132
   **12**  Scatter Plot Command Mode Options  147
   **13**  Setting Text Size  148
   **14**  Scatter Matrices with a Period (.) as the Plotting Symbol  148
   **15**  Restoring the Original Order of Your Data  158
   **16**  Stabilizing UCL and LCL Lines in P Charts  166
   **17**  Additional Options for R Charts and X-Bar Charts  171
   **18**  Warning: SQ Sorts the Data When It Runs the
           Spearman Correlation  233
   **19**  Arranging Your Windows to Display Diagnostic Plots  243
   **20**  AV Plots as an Alternative to Plot Residual versus an X  249
   **21**  The Durbin-Watson Statistic  265
   **22**  Additional Options (Command Mode Only)  285
   **23**  Additional Options (Command Mode Only)  290

# Message from the Stata Corporation

We at StataCorp develop and maintain a package called Stata. Stata is used by professionals for a variety of purposes, ranging from submissions of New Drug Approvals (NDA) to the Food and Drug Administration (FDA) to measuring the employment effects of minimum wages.

Stata is run on machines ranging from supercomputers to laptops, on operating systems including DOS, Windows, Macintosh, and UNIX. Stata is used by researchers at NASA and has been delivered by diplomatic pouch to health researchers in Africa.

Results produced by Stata have been published in professional journals such as the *Journal of the American Medical Association*, the *American Economic Review*, the *American Political Science Review*—just to name a few—and have been presented in courts of law.

Of course, in all these cases, it is the professional researchers—the persons with the statistical training—who did the real work, and Stata was merely a tool they used. Nevertheless, professionals use Stata and we write software for them.

StataQuest is derived from Stata. We added pull-down menus and Windows dialog boxes to make doing simple things easy, and we removed the advanced statistical features. Despite the omissions, the heart of Stata remains, and you can explore it. StataQuest does real statistics.

Stata and StataQuest are not what's important. What's important is that you learn to answer questions for yourself. For instance, you know that smoking causes cancer, but how do you know? You "know" this because you took someone else's word for it. Professional researchers—persons with statistical training—answer questions like this.

To my mind, too much of introductory statistics focuses on artificial questions. What's the height distribution of your fellow students? If the average value of x is 10 in a sample of size 100 and the standard deviation is 3, what is...? My first response, too, is who cares? Given questions like these, it is too easy to lose sight that statistics is about providing real answers to real questions.

Yes, you must learn to crawl before you can walk, but the important thing about the enclosed diskette is that it contains not only software, but more than 60 datasets containing real data. Maybe, as you use StataQuest, this will help you remember that, if you continue, someday somebody will ask you a real question and really care about your answer.

*William Gould*
President, Stata Corporation
College Station, Texas

# Preface for Windows 95

StataQuest has evolved from a DOS-based product to a Macintosh version, to a Windows 3.1 version, and now to the present Windows 95 version. The Windows 95 version incorporates the new features of Windows 95: file names with as many as 255 characters including spaces in them, easier-to-use dialog boxes, and easy switching between programs using the Taskbar at the bottom of the screen.

Similarly, StataQuest's user manuals have evolved. This manual is the StataQuest 4 for Windows and DOS user manual modified for Windows 95. We have included new dialog boxes for opening files, opening a log window, and running a program. We have modified the text for labeling data, working with dialog boxes, switching programs, using long file names, and exiting the program. There are still some references to DOS or Windows 3.1, but these references are in this book only; StataQuest for Windows 95 is a full Windows 95 product and works even better and more smoothly than the Windows 3.1 version.

As before, we think StataQuest for Windows 95 is "just what the doctor ordered" for beginning statistics students.

*J. Theodore Anagnoson*          *Richard E. DeLeon*

# Preface

As teachers of computing, statistics, and data analysis, we have always looked for software that makes instruction easier and learning faster and more fun. We found it in StataQuest.

StataQuest for Windows is a terrific piece of Windows software. All of the procedures you would want to use in a beginning-to-intermediate level statistics class are available right from the menus. It is all point-and-click. There is no layering of Windows dialog boxes on top of each other. StataQuest responds very quickly, provides excellent on-line help, and gives you crisp, accurate, easy-to-interpret output.

StataQuest includes this book, the software, and more than 60 data files on the StataQuest disk. StataQuest has all the standard features you would expect, including a spreadsheet-like Editor to enter data. Its more innovative features include the ability to do do all the basics of data analysis from one sub-menu; a random number generator; robust and logistic regression; a dialog box that makes it easy to do regression diagnostics; graph options for placing regression lines through scatterplots, naming the points in such graphs, and sizing the points by a third variable; and the ability to solve math problems, run statistical tests, and compute confidence intervals right from the menu with minimal typing.

- StataQuest for Windows is all point-and-click menus. It is easy to use and speedy in its execution. You can easily alternate between point-and-click with the menus and command mode. Except for advanced multivariate and factor analysis, StataQuest contains all the basic Stata commands.

■ StataQuest for Windows does all the basic statistical and data analysis tasks that students have in the first two or three courses in statistics, data analysis, or research methods. It includes everything from running t-tests and various nonparametric statistics to multiple, robust, and logistic regression.

■ StataQuest has a strong emphasis on graphs for both data analysis and data presentation. All graphs and statistical results can be printed, inserted into a word processor through the Windows clipboard, or saved for future use.

■ StataQuest will hold 4,000 data points (the number of variables multiplied by the number of observations). Data sets can range in size from as many as 25 variables and 160 observations at one extreme to 6 variables and 600 observations on the other.

■ The StataQuest software comes with this book, which includes how-to instructions for every menu item, guidelines for interpreting the output, and problems at the end of each chapter.

## Our Approach to Using StataQuest as a Teaching Tool

We have a "learning by doing" philosophy of data analysis and statistics. We have included many data sets that can be used for class assignments and exercises. The problems and exercises we have placed at the end of each chapter are based on these data sets and are designed to make it easy for you to learn StataQuest and explore some of its innovative features. Instructors and students can also create their own data sets, either by entering them from the keyboard using StataQuest's spreadsheet, by inputting a text file, or by converting them from other formats with DBMS/COPY or Stat/Transfer. The only limits are the regular StataQuest maximums.

Instructors who take an exploratory data analysis (EDA) approach will find a strong set of EDA tools in StataQuest's box plots, stem-and-leaf plots, and the other graphing features. Those who take a more traditional approach to data analysis will find all their familiar tools available and may enjoy the graphical tools provided.

StataQuest is a subset of professional Stata that has been customized for student use. Both of us have used professional Stata for years in our academic work and teaching. As Stata enthusiasts, we believe StataQuest's easy-to-use menu is "just what the doctor ordered" for bringing Stata's power and speed to beginning students.

# Organization

This book is a user manual for StataQuest. It is organized in six sections:

- **Part I—Getting Started** contains one chapter that tells you how to do all the basics: start and end the program, configure SQ's Windows, place the text and graphs from SQ in your word processor, print output and graphs, and so on. (Inputting data is handled in Chapter 2.)

- **Part II—Data**, contains four chapters. These help you get your data into StataQuest's format (Chapter 2), change your data in any way (Chapter 3), find out what is in a data set (Chapter 4), run basic statistics (Chapter 4), and produce frequency distributions and cross-tabulations (Chapter 5).

- **Part III—Graph** contains seven graph chapters, a major expansion from the previous edition. We now have individual chapters on each of the major graphs that SQ produces: histograms and normal quantile plots (Chapter 6); stem-and-leaf and dot plots (Chapter 7); box plots (Chapter 8); bar charts (Chapter 9); scatter plots (Chapter 10); time series plots (Chapter 11); and quality control charts (Chapter 12).

- **Part IV—Statistical Tests**, contains three chapters, one on SQ's parametric tests (Chapter 13); one on nonparametric tests (Chapter 14); and one on the Statistical Calculator (Chapter 15), that runs parametric tests from summary data entered directly from the keyboard. The Calculator includes a feature called Statistical Tables, in which you can enter the value of a distribution and get back the value of the area under the tail of the distribution. The Calculator includes the same feature in reverse, so that you can input the probability and receive back the value of the normal, Student's t, F, or chi-squared distribution. It includes (of course) both a regular and a RPN calculator.

- **Part V—Correlation, Regression, and ANOVA**, includes not only the normal chapters on correlation (both Pearson and Spearman), simple regression, multiple regression, and ANOVA, but also full chapters on robust regression and logistic regression, as well as a section of the ANOVA chapter on repeated measures ANOVA.

# Acknowledgments

No project as substantial as this one could be accomplished without a lot of help from a lot of people. Bill Gould, James Hardin, Alan Riley, and Bill Scribney of the Stata Corporation wrote and programmed a wonderful piece of software. Stan Loll of Duxbury Press originated the idea of a student menu-driven piece of software and carried it through to fruition and a second edition amidst more obstacles than any of us could have dreamed would occur when we first discussed this project. Curt Hinrichs of Duxbury Press has helped us with StataQuest 4 for Windows 95. Robin Gold of Forbes Mill Press was instrumental in producing a book from our manuscript.

Others we would like to thank include Lawrence C. Hamilton of the University of New Hampshire, Rufus P. Browning of San Francisco State University, Pat Branton of the Stata Corporation, John Allswang of California State University, Los Angeles, Arlene K. DeLeon, John Hsu, Barry Hill, the now more than 100 faculty from across the nation who have attended any of the five workshops we have held on exploratory data analysis using micro-computers, the National Science Foundation (which supported four of those workshops), and the California State University's Social Science Research and Instructional Council (SSRIC), whose advice and stimulation encouraged both of us to pursue new ways to analyze data and look at research and instructional problems. None of these people, however, is responsible for any errors and omissions.

## Dedication

To Louise and Arlene, with our love and thanks

**J. Theodore Anagnoson**
California State University,
Los Angeles, CA 90032-8226
anagnosn@vais.net

**Richard E. DeLeon**
San Francisco State University
San Francisco, CA 94132
rdeleon@sfsu.edu

# *Getting Started*

## What Is StataQuest?

StataQuest is a powerful and easy-to-use microcomputer software package for quantitative data analysis. StataQuest's statistical procedures and graphical displays will enable you to describe and explore a set of data, discover its patterns and relationships, and test hypotheses. Even the most brilliantly designed and flawlessly executed research yields little of value if the data collected aren't analyzed with skill and the proper tools. In this book, we discuss each of StataQuest's tools, show you step-by-step when and how to use them, demonstrate their use on actual data sets, and offer suggestions for interpreting the output. This chapter tells you how to install SQ, how to configure SQ so that its windows meet your needs and preferences, what SQ's basic rules are, and how to use SQ's several utilities.

## Requirements and Limits

StataQuest's system requirements are as follows:

- StataQuest for Windows 95 system requirements are any DOS computer that will run Windows 95, has a 3.5-inch floppy disk drive, at least

| | |
|---|---|
| ■ ■ ■ ■ ■ ■ ■ **IN THIS CHAPTER** | ■ **Requirements and limits** |
| | ■ **Installing, starting, and stopping StataQuest** |
| | ■ **Configuring StataQuest to your preferences** |
| | ■ **Basic Rules—Windows** |
| | ■ **Basic Rules—DOS** |
| | ■ **Getting help** |
| | ■ **How to place Results window text in your word processor** |
| | ■ **How to print your work** |
| | ■ **Printing StataQuest graphs** |
| | ■ **Telling the difference among StataQuest's files** |
| | ■ **StataQuest's utilities** |
| | ■ **Structure of a typical StataQuest session** |
| | ■ **What's really happening when you choose commands from StataQuest's menus?** |
| | ■ **What's next?** |

enough RAM to run Windows, and any printer that can be configured for Windows. StataQuest requires approximately 1 megabyte of disk space to store its programs and data files. No additional memory is necessary beyond the minimum required to run Windows 95.

## StataQuest's Maximum File Size

Each row of your data set represents a **case** or **observation** (we use these terms interchangeably). These are the units of analysis in research and can be, for example, cities, individuals, years, circuit boards, planets, wars, birds, and so on.

Each column is your data set represents a **variable**, an attribute or characteristic of each case that varies from case to case. If there is no variation, an attribute is a **constant** for all cases studied (if your data set is all males, then the variable "gender" is a constant). The **values** of each variable are the possible attributes a case can assume. Individuals classified on the variable "gender," for example, can assume the values of "male" or "female." National populations measured on the variable "average life expectancy at birth" can assume continuous values in the realistic range of 30 to 80 years. Students measured on the variable "letter grade" might be assigned the values of A, B, C, D, or F. This last example is an instance of a special kind of variable called a **string variable**, which is any combination of both letters and numbers used to identify and differentiate cases but possessing no mathematical properties. Mathematical operations such as addition and subtraction cannot be performed on string variables. (In SQ, if a variable is not explicitly identified as a string variable, it is assumed to be numeric.)

**StataQuest's maximum file size is 4,000 data elements.** These can represent a maximum of 25 variables (columns), in which case you can have as many as 160 rows or observations. Or you can have as many as 600 observations, in which case you can have a maximum of 6 variables. The total number of data elements is the product of the number of variables times the number of observations, or rows x columns.

**You can work without data in memory.** StataQuest's Calculator menu performs one- and two-sample tests of the differences between means (both for large and small samples [t-tests]), tests of the differences between proportions, a standard deviation test, a confidence interval, binomial tests, and Poisson tests. In addition, you can find the significance level of any t-value, chi-squared value, or other statistical distributions, and the calculator menu has both a regular calculator and one with Reverse Polish Notation for those used to Hewlett Packard scientific calculators. All of these functions are covered in Chapter 15, *The Statistical Calculator.*

# Installing, Starting, and Stopping StataQuest

The directions for installing StataQuest are different for the Windows and DOS versions.

## Windows 95

### *To install SQ*

1   Make sure that Windows 95 is installed on your computer. Start Windows. From the Taskbar, click on *Start* and then on *Run*.

2   Place your SQ disk (or disk 1 if you have a 2-disk set) in either the a: or b: floppy disk drive.

3   In the dialog box, type:  a:setup (or b:setup if your SQ disk is in the b:drive). See Figure 1.1. Then click on "OK."

SQ for Windows by default places its data files (which have the tail .dta) in a subdirectory called *c:\sqdata*; its program files will be placed in the *c:\wstataq* subdirectory.

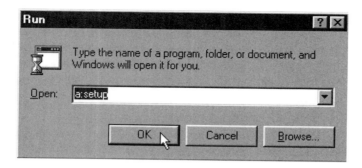

**Figure 1.1** *Windows 95 File / Run Dialog Box*

### *To start SQ*

Start Windows if it is not already running. Then do one of the following:

- Double-click (two quick clicks on the left mouse button while pointing at the SQ icon in Windows) on the SQ icon
- Choose StataQuest from the Taskbar.

### *To exit from SQ*

From the *File* menu, choose *Exit*. If you have changed the data, SQ will tell you that the data set has changed and ask if you really want to exit. You must answer OK to exit.

## DOS

### *To install SQ*

Insert the SQ disk into a floppy disk drive of the appropriate size, make sure that drive is the current drive (example: type a: and press ENTER), type go, and press ENTER. The StataQuest Utility menu has an item that allows you to install SQ to your hard drive.

**Note:** If you are running SQ directly off the SQ floppy disk (as opposed to saving your data files on a floppy disk), you do not need to install SQ. The installation procedure is for users with hard disks.

### *To start SQ*

Choose the top item (*Start SQ*) from the StataQuest Utility menu.

***To exit from SQ***

From the *File* menu, choose *Exit*. If there are data in memory, SQ will ask you whether you want to save the data, even if the data have not changed. Choose Quit from the *StataQuest Utility* menu to exit to DOS

# Configuring StataQuest to Your Preferences

StataQuest for DOS has a default full-screen configuration, but users of SQ for Windows can change the location and size of each of SQ's Windows.

## Windows

SQ comes with four default windows. Beginning users will generally use a mouse or ALT+a_key (hold the ALT key down and press another key once; the other key is underlined in the menus) to choose the command or program they want to run. Their results will appear in the **Results window**.

Intermediate and advanced users will choose their commands and programs in one of the following ways:

■ With a mouse.

■ With ALT+a_key.

■ With the Command window, typing their commands and options.

■ With the Command window, bringing up previous commands with PGUP or by clicking on the previous command in the Review window. Intermediate and advanced users sometimes add variable names to a command in the Command window by clicking on a variable name in the Variables window.

Thus,

■ **Users who choose their commands from the menus** will use the Results window most of the time. Even those who rarely use command mode, however, may find it useful to run commands on subsets of data (see Chapter 3, *Modifying Your Data*) using the **Command** and **Review** windows.

■ **Users who use command mode** will type their commands in the Command window, obtain previous commands with the PGUP key or by clicking on them in the Review window, and obtain variable names without having to type them by clicking on them in the **Variables window**.

**Note:**

■ **Once you close the Review window**, you can't get it back during the same session unless you go back to SQ's default windowing configuration by choosing *Default Windowing* from the *Prefs* menu.

- **You can move a window to a different location** by holding the left-mouse button down when pointing to the "title bar," the shaded rectangle at the top of a window with the window title in it, and moving the mouse.
- **You can change the size of a window** by moving the mouse to the margin you want to change, where its shape will change to a two-headed arrow (↔). You then hold the left mouse button down, move the margin to the desired location, and release the mouse button.
- **You cannot close the Results window or the Command window**, although you can cover the Command window with the Results window if you want to (we don't recommend this action, as the Command window will reappear after file opening and saving operations).

We will use these rules to configure SQ for Windows to our liking. When you start SQ for the first time, you should see a windowing configuration that looks something like Figure 1.2.

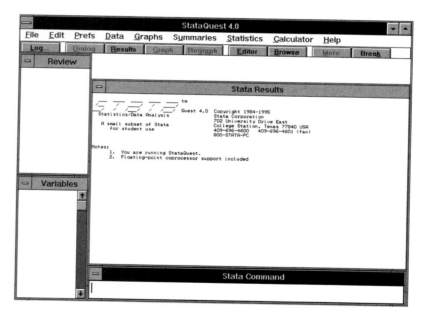

**Figure 1.2** *StataQuest Default Windows*

First, use the mouse to raise the top border of the Results window until it begins just below the **Editor** button. (Move the mouse to the top border until the cursor changes to a vertical two-headed arrow; then hold the left mouse button down and move the border upwards.) The result is shown in Figure 1.3.

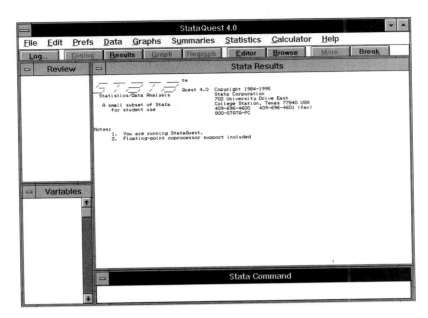

**Figure 1.3** *StataQuest Windows After Enlarging the Results Window*

If the Results window is hard to read, use the mouse to select the button in the upper-left corner of the Results window. A special menu will appear in which the bottom item is *Fonts*. Selecting *Fonts* allows you to **change the default font** in the Results window. **Caution:** If you change the font to one that is too large, some of SQ's output will appear off the screen to the right.

You can experiment with different fonts in the Results window. We recommend, if your screen is large enough, the font labeled "Stata 8x12," which has a 9 point type size instead of the default 6 point. With a 17" monitor, this makes an attractive window, as shown in Figure 1.4. Most of the examples in this book will come from windows that look like Figure 1.4.

It is also possible to enlarge the Results window so that it takes up most of the screen, as we have done in Figure 1.5. The choice is a matter of individual preference, the size of your monitor, and the degree of resolution of your graphics card. Try different arrangements, and use the one that you like the best.

When you find the one you like, save it so that SQ will begin with it the next time you start SQ. To save it, from the *Prefs* menu, select *Save Windowing Preferences*. This option will save your present window configuration, and SQ will load next time with this preference. **Caution:** If you are using a networked lab, you cannot save your windowing preferences. You must configure your windows each time you load SQ.

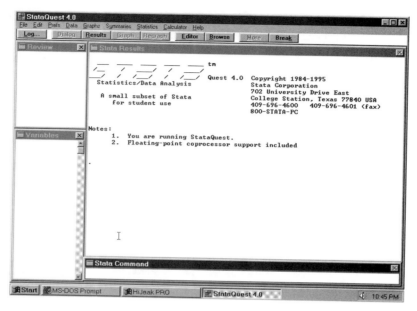

**Figure 1.4** *StataQuest Results Window with 8x12 font*

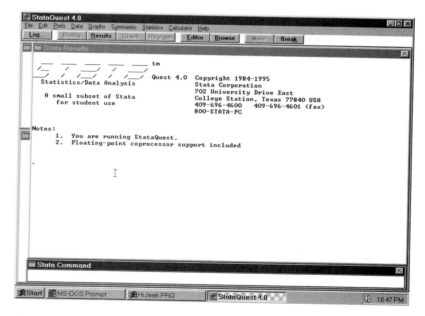

**Figure 1.5** *Results Window Enlarged to Take Up Most of the Screen (8x12 font)*

# Basic Rules—Windows

## How to Work with Windows

SQ opens with four windows showing, the Results window, the Command window, the Variables window, and the Review window. If you run a statistics command, a dialog box will open, and on some statistics commands, the dialog box will stay active until you close it. If a dialog box is active, the **Dialog** button will be highlighted. Similarly, most of SQ's graph commands will run with the *Draw* option, which leaves the dialog box active so that you can change one of the graph command's options or variable names and run the graph again to see what difference the change makes. Here, too, the Dialog button will be highlighted so you can see that the dialog box is active and must be closed or finished with before you can proceed.

In the dialog box for a graph command, the *Draw* option executes the command, and the dialog box stays active. The *OK* option executes the command, and the dialog box disappears.

When a dialog box is active, the *Data*, *Graphs*, *Summaries*, *Statistics*, *Calculator*, and *Help* menu commands are "blanked out" at the top; they cannot be chosen until the dialog box has been closed. If the Results or Graph window covers a dialog box, you can bring the dialog box to the surface by clicking on the Dialog button.

**How to bring a window to the top.** If windows overlap, just click on the part of one that is showing to bring it to the surface. Or use the buttons on the button bar at the top of the screen.

**How to close a window.** You can close any window (other than the Command window and the Results window) by clicking on the button in the upper-left corner of the window. This activates a small menu, the last or second-to-last item of which is *Close*. Windows 95 also has an X button (**x**) in the top right corner that will close any window.

**How to change the font size of windows.** You can change the fonts of the following windows: Results, Editor, Review, Variables, and Help. To do this, click on the button in the upper-left corner of the window. The last item on the small menu should be *Fonts*. You cannot change the font in the Command window, the Log window, or the Graph window.

**How to move a window.** You can move windows around the screen, although we have not found that moving anything other than the Results window improves the usability of the SQ screen. You can move a window by clicking with the left mouse button while pointing to the title bar of the window. Hold the mouse button down and move the window to the desired location.

**How to change the size of a window.** Move the mouse to edge of the window you desire to change. Hold the left mouse button down when the mouse pointer changes to a two headed arrow (for example, ↔), and move the edge to the desired location.

**How to choose commands.** Normally you will choose a command by pointing to the menu at the top of the SQ screen. Some users may want to type in commands in the Command window. Any command that has already been executed can be brought back for execution by pointing to it in the Review window. It can then be edited in the Command window before execution. Any command in the Command window can be executed by pressing the ENTER key.

## How to Work with Dialog Boxes

When you choose a command from the top menu, SQ will present a dialog box. These fall into two categories. Figure 1.6 shows a dialog box used to select files, in this case to open a file and read it from disk to memory.

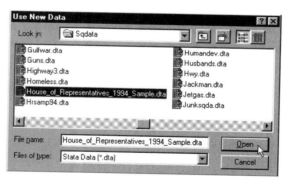

**Figure 1.6** *The Use New Data Dialog Box*

In the first row of the dialog box, you can "**Look in:**" either the Sqdata subdirectory or some other subdirectory. If you hold the mouse cursor over any Windows icon, the default action produces a small box with the name of the procedure in it. The icon with the up arrow in it allows you easily to move up one level in your subdirectories. The next icon allows you to open a new folder at the level indicated in the "Look in:" box. The next icon allows you to choose between displaying a "list" of files, as we have done above, or "details" about those files, which produces the usual DOS/Windows directory listing of file name, date, size, and so on.

The large box in the middle has a list of files. You scroll by using the horizontal scroll bar at the bottom. The "**File name:**" box allows you to screen the files by displaying a subset of all possible files. To display files beginning with the term "air," for example, change the "*.dta" in the box to "air*.dta." The "**Files of type:**" box at the bottom allows you to display either all files, or the default, StataQuest's *.dta files.

A second type of dialog box requires you to select variables. Figure 1.7 shows the box for SQ's Scatterplot Matrix command, on the Graphs menu.

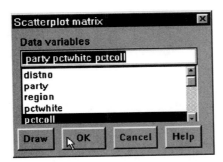

**Figure 1.7** *The Scatterplot Matrix Dialog Box*

You can select variables by typing them in or by clicking on them with the left-mouse button from the list provided. You can scroll the list by clicking on the up arrow, down arrow, or within the gray box connecting them.

To remove a variable once you have selected it, either double-click on its name and press the delete key ("Del" or "Delete"), or insert the cursor after the variable name and backspace until it is removed.

On graph commands, there are Draw and OK options. The Draw option executes the graph command but leaves the dialog box active on the screen so that you can change the parameters of the graph by, for example, changing the variable list. The OK option executes the command and closes the dialog box.

**Note:** If a dialog box is still active, the Dialog button on the second row at the top will be highlighted, and other SQ statistical commands will be "whited-out" and cannot be chosen until the dialog box is closed.

## How to Work with Menu Items and Buttons

What do the menu items and buttons mean at the top of the SQ for Windows screen? (You can click on each of these to see the menus as you read about them.)

### Menu items:

- **File.** Opens and saves SQ data files, saves or prints graphs (if a graph window is open), prints a log file if one is active, imports and exports plain character files, and exits SQ.
- **Edit.** Allows you to copy output from the Results or Graph windows to your word processor or other application.
- **Prefs.** Allows you to go back to SQ's default windowing or your own option as saved previously: Editor, Graph (line thickness, colors), Results window (colors). **Note:** You change fonts in the Editor and Results window through the button in the upper left of the respective window.
- **Data.** Options to generate new variables; to replace values in old variables; to change labels to strings and vice versa; to label a dataset, variables, or values; and to enter the Editor.
- **Graphs.** Chooses all of SQ's graph types except those which are for regression or ANOVA diagnostics; the latter are on pop-up menus that appear after the regression or ANOVA.
- **Summaries.** Lists data or variables, and generates means, standard deviations, and other single-variable statistics for the data set as a whole or subsets. The Tables sub-command produces frequency distributions, and cross-tabulations.
- **Statistics.** Produces all of SQ's parametric and nonparametric statistics on data in memory, also correlation, regression (simple, multiple, logistic, robust), analysis of variance.
- **Calculator.** Does calculations without data in memory. Produces t-tests, z-tests, binomial and Poisson tests, has statistical tables to look up any z-value, t-value, chi square, and so on, plus regular and Reverse Polish Notation calculator.
- **Help.** Gives you SQ's on-line help for its commands.

### Buttons:

- **Log.** Makes a record of your SQ session that you can print out during your session or after. See page 15.
- **Dialog.** Brings the active dialog box to the top of the screen.
- **Results.** Makes the Results window the top-most window.
- **Graph.** Brings the open graph window to the top of the screen.
- **Regraph.** Executes the most recent graph command with the current data.
- **Editor.** Makes the spreadsheet editor the active window. When the Editor is active, all other menus are inactive.

- **Browse.** Makes the spreadsheet editor the active window, but does not allow you to change the data. Use it for looking at the data without the danger of accidentally changing anything.
- **More.** If the printout from a command will take more than one screen in the Results window, a "—more—" will appear at the bottom of the screen. You can clear the —more— by pressing any key, or, if you are using a mouse at the time, clicking on the More button.
- **Break.** An alternative for CNTRL-BREAK at the keyboard. Stops SQ from whatever it is doing so that you can issue another command. Will **not** throw you out of SQ.

## Basic Rules—DOS

- You access the menus with ESC and the arrow keys.
- The menu outline is similar to SQ for Windows except that *Edit / Spreadsheet* accesses the editor, and new variables are generated inside the Editor. All labeling is also done inside the Editor.
- You always press ENTER after choosing a command.
- Pressing ESC takes you one level up in the menus.
- Pressing the F10 key in the spreadsheet takes you to the menus.
- Pressing F1 accesses the Help system.
- Pressing F5 turns a log file off and on once the log file has been established with the *Files / Session logging* command.
- To choose an item off the menu, use the right, left, up, and down arrow keys to go to the item, and then press ENTER.
- If SQ exits you from the menu system—you will know that this has happened if you are facing a dot (._ ) prompt—you can reenter the menu system by typing m and pressing ENTER.
- If SQ has more than one screen of information, it will prompt you with a —more— at the bottom of the screen. Pressing any key will result in the next screen or the next menu.
- When SQ has finished a procedure and is about to return to the menu, you will see *Press any key to continue* at the bottom of the screen.

## Getting Help

- **Windows:** Choose *Help* from the top menu bar.
- **DOS:** Press F1 to access *Help*.

# How to Place Results Window Text and Graphs in Your Word Processor

## Results Window Text

Use the mouse to select the text in the Results Window. Then, from the *Edit* menu, choose *Copy Text*. Use the *Taskbar* at the bottom of the Windows 95 screen to switch to your word processing program (or you can use the Windows command ALT-TAB). Then, from the *Edit* menu, choose *Paste* to insert the text to the current cursor location. Use a plain typewriter font such as Courier, size 9 or 10 point, not a proportional font like Times Roman or Arial on your text or table.

## Graphs

After running the graph, choose *Copy Graph* from the *Edit* menu. Use the *Taskbar* at the bottom of the Windows 95 screen (or the Windows command ALT-TAB) to shift to your word-processing program. Then from the *Edit* menu, choose *Paste* to insert the graph at the current cursor location. In some word processors, the graph can be reduced to size by inserting it into a frame of the appropriate size. Check your word processor help files or manuals for more information.

# How to Print Your Work

A SQ log file is a file that contains the commands you generate through choosing SQ menu items as well as your results. The most convenient way to print your results is to establish a log file of your session, saving and printing it at your convenience.

Use the mouse to click on the Log button or CTRL-L if you are using the keyboard. An Open Stata Log dialog box appears, with *.log highlighted and any other log files in the \sqdata subdirectory listed. Provide a name for the present log and choose OK. We suggest that you date your log files by choosing names like "log618" for the log file of June 18th or some such similar system. If you call your log file "log618," SQ will form a file called *log618.log* on your default drive. With Windows 95, you can give logs long names such as "Log for Professor DeLeon 8-22-96" and the like. Note that you cannot use a slash in a Windows 95 long file name, but dashes are OK. Thus, "Log 8-22-96" is OK, but "Log 8/22/96" is not.

If the log file begins successfully, SQ will place the log file on top. Click on any portion of the Results window to cover it up.

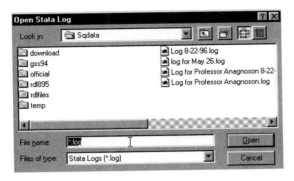

**Figure 1.8** *Open Stata Log Dialog Box*

If you again choose the Log button, you can do any of the following:

- Open a Log window that allows you to look at the log file
- Suspend the log file
- Close the log file

To print the current log file, choose *Print log* from the SQ *File* menu. Click on OK to print the entire current log file.

Note that *File / Print log* prints the current log file only. To print an old log file, choose the Log button, select the old file from among those listed (or change disk or subdirectory to find it and then select it) and click on OK. Then, from the *File* menu, choose *Print* log.

# Printing StataQuest Graphs

Note in all versions of StataQuest that all graphs except dot plots and stem-and-leaf plots are saved in special graph files, not in the log files that you otherwise use to print out your output.

## Windows

The procedure in general is to choose a graph command, produce the graph, and either to save it or print it when the Graph window is active. Graphs can be saved in "Windows metafile format" (*.wmf) or in StataQuest's own format

(*.gph). The "Windows metafile format" can be transferred through Windows into other programs, for example, your word processor. The StataQuest format is used for graphs you may want to look at in the future inside StataQuest using the *View saved graphs* command from the *Graphs* menu. In general, most if not all graphs should be saved in the Windows metafile format. Thus,

1 To save a graph: From the *File* menu, choose *Save Graph*. The default save format is a Windows metafile (*.wmf). You can also save a graph as a StataQuest graph by choosing the "Stata graph file" format (*.gph). Give the graph a name using one to eight characters.

2 To print a graph: From the *File* menu, choose *Print graph*. The graph will print on whatever printer is configured for Windows. You do not need to save a graph first to print it.

## DOS

With the DOS version of SQ, you produce and save the graph in SQ, and you print it and any other saved graph from the StataQuest Utility menu once you exit SQ.

To print a graph file, first form the graph. After the graph displays on the screen, SQ will ask you whether you want to *Continue, Show graph again*, or *Save graph for printing*. Choose *Save graph for printing* and give the graph a name of eight characters or less.

Then after you exit SQ, you choose *Print a SQ graph* from the *StataQuest Utility* menu and select a graph for printing. SQ will then ask whether it should print in HP LaserJet format (for laser printers) or Epson FX format (for dot matrix printers). Choose either, and the graph will print on your printer. If the cursor "hangs" after the printing is finished, press ENTER.

# Telling the Difference Among StataQuest's Files

In the course of a typical session, you will work with and produce a number of different files. DOS has a standard file format: as many as eight characters for a name and as many as three for a tail, with a period between them. Thus: *log62495.log; restaur.dta; test1.gph; mydata.raw*. Here is how to tell the difference among them:

■ **Raw data files.** These are formed with an editor or word processor, or are raw data outputted from another program. You can look at any raw data file with the DOS `type` command, or with any editor or word processor. A file can have any name within the standard DOS structure of an eight-character name and three-character extension, but SQ assumes (if you don't tell it otherwise) that a raw data file being read in to SQ's format through the *Import ASCII* command has the tail *.raw* or *.dat*.

- **StataQuest data files.** These are formed when you choose *Save* or *Save As* from the *File* menu and save the file in SQ's own special, efficient format. These always have the tail ".dta" and can never be looked at with an editor or word processor. If you do, you will see gibberish, that is, special symbols.
- **Log files.** These are formed when you press the Log button. They, like raw data files, can be read with a word processor or editor. You can print them from your word processor, from within SQ (from the *File* menu, select *Print log* to print the current log), or with the DOS print command. They all contain the tail ".log" plus whatever name you gave them when they were established.
- **StataQuest graph files.** These are formed within StataQuest when you choose *Save graph* from the *File* menu. This command saves the current graph for later printing. Graph files, like StataQuest data files, are formed with special characters in an efficient format and contain complicated printer instructions to tell the printer how to draw the graph. You cannot look at these with your word processor or editor. They contain the tail ".gph" or ".wmf." Files with the tail ".wmf" can be placed in a word processing file in Windows.

## StataQuest's Utilities

The DOS version of SQ has several utility programs for handling file functions within the program, but in the Windows version of SQ, the utility functions are handled by Windows utilities.

### DOS

The DOS version of SQ has a *Maintenance* sub-menu under *File* with three "housekeeping" commands. The *Directory* command lists the names of all files contained in the current folder or directory, along with the file size. This information could be important if you are short on space for saving new files. The *Display* command displays the contents of any text file (text files are raw data files or log files; see the previous section) in the current folder or directory. This command can be used to look at log files or data files in text format before they are read into SQ. (It cannot be used to look at StataQuest's ".dta" files, which are saved in a special format.) The *Erase* command erases a file from the current folder or directory.

## Structure of a Typical StataQuest Session

You have now had an introduction to all the basics of a StataQuest session. Here is a review of what gets done in what approximate order:

1. **Start** StataQuest.
2. Either:
    - Run problems on the **Statistical Calculator** which does not require data in memory
    - **Enter data** through the Editor (see Chapter 2)
    - **Open a data file** and read it into memory with *File / Open*
3. **Open a log file** to record your results for later printing.
4. **Perform analysis** as needed.
5. **Create any necessary graphs**, printing as you proceed.
6. **Print graphs** if desired. (**DOS:** do this from the StataQuest Utility menu after exiting.)
7. **Print log file** if desired. (**DOS:** do this from the StataQuest Utility menu after exiting.)
8. **Save data file for later use** if it has changed during the current session.
9. **Exit** StataQuest.

## What Is Really Happening When You Choose Commands from StataQuest's Menus?

Here's what's happening inside your computer. There is a statistical package called **Stata**, produced by the folks at Stata, Inc. in College Station, Texas. **Stata** is command driven, that is, you type a command like `describe` or `summarize var1, detail`, and you receive your results. It is interactive, which means you get your results back immediately after you issue your command; you study the results, then you issue a slightly revised command, and so on.

**StataQuest** is a menu-driven "front-end" for Stata. When you choose a StataQuest menu item, you are choosing to issue a command in Stata's language. In fact, in both the DOS and the Windows versions of StataQuest, Stata prints its command right on the screen above the results. You can even reissue the command (by recalling it with PGUP and then altering it if you desire) through the Command window (Windows). Stata is a full-fledged statistical package that is used in academia, government, and the private sector to analyze data; StataQuest is a limited subset of the Stata package developed especially for students and available through Duxbury Press at a special student price.

If you want more information on the statistical package Stata, you can find more information in Stata's Preface to this book and from the *About StataQuest* menu item under *Help*.

## What's Next?

At this point, you have a choice. You can go directly to Chapter 2, *Entering Data into StataQuest*, to learn how to input data. Chapter 2 explains how to input your data through the data editor, through external text (ASCII) files, or through a utility program to transfer statistical package files from one program to another. Then the rest of the book, except for the chapter on the statistical calculator, explains the different procedures you can run with data in memory.

Alternatively, you can go directly to Chapter 15, *The Statistical Calculator*, to run statistical tests directly with statistics (like the mean, standard deviation, and sample size) that you input without data being in memory.

## ■ EXERCISES

**1.1** Install StataQuest on your own computer if you have one.

**1.2** Configure StataQuest's main windows according to your preferences as per pages 4–8. Save your new windowing configurations with the *Save Windowing Preferences* command from the *Prefs* menu.

If you are running SQ in a lab, decide what windowing configuration you prefer and how to create it each time you run SQ.

**1.3** SQ comes with two graphs that you can print. In SQ for Windows, use the *View saved graphs* command from the *Graphs* menu to make one of these active and print it. Do the same with the log file that comes with SQ using the Log button in Windows.

# 2

# *Entering Data into StataQuest*

## The StataQuest Editor

StataQuest's spreadsheet, which we call the Editor, is a powerful tool that allows you to enter, edit, and analyze data. With it, you can enter a new data set, browse an existing data set, or make changes to a data set. In addition, you can add variable labels to your data set, identifying each variable name by a descriptive label of as many as 31 characters.

Be sure you differentiate between StataQuest's own data files and data that you've obtained from someplace else in a plain ASCII file.

- **A StataQuest data set** is in SQ's own format, either inputted through the spreadsheet, from an ASCII file, or converted from another statistical package or spreadsheet. SQ data sets are saved with a file name in Windows 95 of 255 characters or less plus ".dta" as the tail. SQ data sets are saved in a special, efficient format.

- **An ASCII file** is a file that you can look at with your word processor or with the DOS type or more commands. It consists of plain characters only. SQ may be able to read this file, but it is not in SQ's own format until you have SQ read it using the *Import ASCII* command from the *File* menu.

■ ■ ■ ■ ■ ■ ■

## IN THIS CHAPTER

- ■ Inputting your own data through the SQ Editor
- ■ Inputting your own data through external (text) files
- ■ Creating random variables
- ■ Stata files
- ■ Spreadsheet and statistical package files
- ■ Labeling your data
- ■ Renaming variables
- ■ Sorting
- ■ Exiting the Editor

# Inputting Your Own Data Through the StataQuest Editor

All spreadsheets work with **cells**, a cell being the location where you put one data point. A **data point** is a single unit of information on a single observation. For example:

- ■ If your observations are people, a single data point might be someone's age.
- ■ If your observations are counties, a single data point might be the number of people in one county.
- ■ If your observations are nations, a single data point might be one country's Gross National Product.

Each cell is denoted by its row number and column label. In StataQuest, **rows** begin at 1 and continue to the bottom of the data set, subject to the overall limit of 600 observations.

Each **column** is a variable that has the same meaning for each observation. Thus, for a data set of countries, a column that is headed by *gnp* will have the Gross National Product for each country in that column. For a data set of people, a column that is headed by *age* will have each person's age consistently coded. That is, SQ assumes that each person's age is coded in years or months or days or some other unit, but always the **same** unit. There is a limit of 25 columns or variables in SQ.

▪ **TIP 1** ▪ ▪ ▪ ▪ ▪ ▪ ▪ ▪ ▪ ▪ ▪ ▪ ▪ ▪ ▪

### Data in Memory versus Data on Disk

- You should always save any changes you make to an existing SQ data set to disk. Remember that SQ changes the data in memory and that none of these changes will be available for you to use the next time you use SQ unless you save them to disk.
- Within the Editor, making changes to the data changes only the data in memory. Leaving the Editor leaves these changes intact in memory. You must save the data set on disk through the *Save* or *Save As* commands from the *File* menu to preserve these changes for a subsequent SQ session.

## The Windows Editor

**To start the editor,** click on the **Editor** button with the mouse, type `edit` in the Command window, or, from the *Data* menu, choose *Edit data*. Notice that the spreadsheet opens on top of whatever other SQ work you have been doing, and that the SQ main menu bars stay on the screen. You can adjust the size of the spreadsheet window in the same way that you configured SQ's main windows in Chapter 1: by holding the mouse button down once the cursor changes from a vertical line to a two-headed arrow (ex: ↔).

The first, second, and third rows are the standard SQ title bar, menus, and buttons. The next row is SQ's Editor title bar, followed by buttons that work in the spreadsheet and a list of variable names.

### Entering Data with the Windows Editor

The basic procedure with the Windows Editor is to decide first whether you want to enter the data across each row or down each observation:

- If you want to go across each row, you enter each cell and press TAB to go to the next cell **across**.
- If you want to go down each column, you enter each cell and press ENTER to go to the next cell **down**.

After the first row or column, SQ will remember how many variables or observations are in the dataset and automatically go to the next observation or variable.

If you start with an empty dataset in memory and open the Editor by pressing on the Editor button or choosing *Edit data* from the *Data* menu, then

**Table 2.1** *Government Spending as a Percentage of GNP (Source: Dye, 1990, 36)*

| Year | Total Public Sector | Federal Government | State-Local Government | State Government | Local Government |
|------|---------------------|--------------------|------------------------|-----------------|------------------|
| 1929 | 10.0% | 2.5% | 7.5% | 2.1% | 5.4% |
| 1949 | 23.0 | 15.3 | 7.8 | 3.0 | 4.8 |
| 1959 | 26.6 | 17.1 | 9.5 | 3.4 | 6.1 |
| 1969 | 30.1 | 17.7 | 12.4 | 4.5 | 7.9 |
| 1979 | 30.6 | 17.6 | 13.1 | 5.0 | 8.1 |
| 1985 | 35.1 | 22.2 | 12.9 | 5.2 | 7.7 |

the editor will open with one line listing the eight buttons, and the second line will read as follows:

```
var1[1]  =
```

This means that the cell on which the cursor is located is variable 1, row 1. If you have data entered in the cell, SQ will list the data to the right of the equal sign.

Let's enter a small data set. Table 2.1 contains data on the percentage of the Gross National Product that government spending forms in the United States. The observations are years. Eventually we will make a graph out of the data and examine the trend to see which level of government is increasing most steeply and during what period the increase is taking place. Before we do that, though, we need to input the data into SQ's own format so that SQ can read the data. We decide to input the data down each column and begin in the top-left cell, cell x1[1]. Here we enter 1929 and press ENTER. The cursor moves automatically to cell x1[2], where we enter 1949. At this stage the basic technique is to enter the number and press ENTER to go to the next cell down.

Notice that

■ We enter each number without the percent sign.

■ We enter numbers without commas or dollar signs ($). For example, we enter 6123456, not 6,123,456 or $6,123,456.

■ Regardless of whether we enter a number like 3.0 with the decimal point and trailing zero, SQ writes the number into the spreadsheet as 3 without the decimal point and trailing zero.

■ If you enter an extra row or column, place the cursor in one cell in that row or column, click on the **Delete** button, and click on the appropriate option, either deleting a row (observation) or a column (variable).

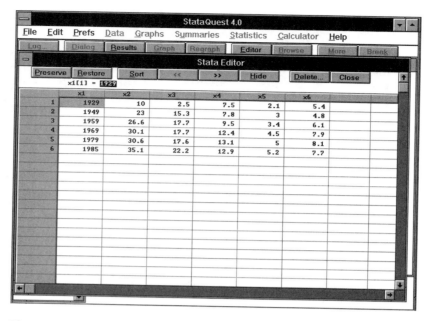

**Figure 2.1** *StataQuest's Editor*

The result should be an Editor window that looks like Figure 2.1. At this point, exit from the Editor and save the file.

**1** Exit the Editor by clicking on the **Close** button.

**2** From the *File* menu, choose *Save*.

**3** SQ will ask you for a name. SQ file names are eight characters or less and composed of letters, numbers, and the underline ( _ ). Enter an appropriate name. If you are doing assignments from your statistics textbook or this book, it is often useful to save files with the number of the assignment in them. Thus the file for the first assignment at the end of this chapter might be called *adc2p1*, "*ad*" standing for Anagnoson and DeLeon, "*c2*" standing for Chapter 2, and "*p1*" standing for problem 1.

**4** Press ENTER. SQ saves your file with the full file name of, in our example, *adc2p1.dta*.

The Editor buttons are for more advanced editing. They do the following:

■ **Preserve.** When you enter the Editor, SQ makes a backup copy of the existing dataset. If you make a series of changes in the data and decide that you are satisfied with these changes, but want to continue editing, you can click the **Preserve** button and do the equivalent of exiting the Editor and then re-entering. These changes change the data in memory, not the data set saved on disk, if any. Thus, when you work in the Editor with an existing data set, you have 3 copies of that data set:

1. The copy that you loaded from disk with the *Open* command from the *File* menu.

2. The copy in memory as you're working in StataQuest.

3. The copy the Editor works with until you tell the Editor that you want your changes to be incorporated into the copy in memory (by choosing Preserve from within the Editor or by exiting from the Editor and keeping your changes). However, these changes are not incorporated into the copy on disk until you *Save* the file.

- **Restore.** When you enter the Editor, a backup copy is made of the existing dataset. If you make a series of changes in the data and decide that you do NOT like these changes, Restore places the backup dataset copy in the Editor.

- **Sort.** SQ sorts the data in the order of the variable where the cursor is located.

- **<<.** SQ shifts the current variable (where the cursor is located) to be the first variable in the data and reorders the data in this order.

- **>>.** SQ shifts the current variable (where the cursor is located) to be the last variable in the data and reorders the data in this order.

- **Hide.** SQ hides the column/variable where the cursor is located so that you cannot see it in the Editor, a useful technique for editing variables. The column or variable still exists, however.

- **Delete.** A pop-up menu asks whether you want to (1) delete the current variable (column); (2) delete the current observation (row); (3) delete all observations where the variable equals the value where the cursor is located. If the cursor is on the *gender* column and on a cell where *gender* = 1, then you would be deleting all observations where gender equals 1.

- **Close.** This button exits the Editor. When you exit the Editor and have changed the data, a dialog box appears stating "Exit Editor: OK to accept changes. Cancel to Restore last Preserve." You can then click on the "OK" button or the "Cancel" button. Clicking on the "OK" button means that the changes made in the Editor should be kept in memory. Clicking on "Cancel" means that the most recent changes, those made in the Editor during the Editing session (or since you last clicked on the Preserve button), are discarded. See the Preserve and Restore button explanations in this section.

## The DOS Editor

Figure 2.2 shows a sample DOS StataQuest spreadsheet.

- Notice that the SQ main menu is still on the screen, at the top. The spreadsheet menu is just below.
- The third line of text tells you the cell SQ is pointing to, in this case, the 1st row and the column headed x1.
- **Enter new value** indicates that SQ is waiting for you to enter more data.

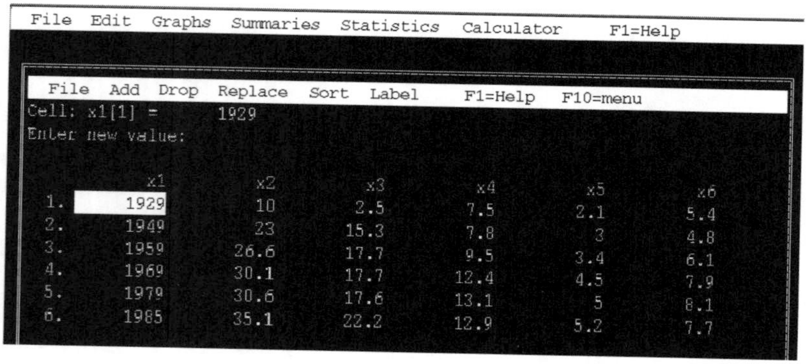

```
File  Edit  Graphs  Summaries  Statistics  Calculator    F1=Help


  File   Add  Drop  Replace  Sort  Label    F1=Help   F10=menu
Cell: x1[1] =        1929
Enter new value:


         x1          x2        x3        x4        x5        x6
  1.     1929         10       2.5       7.5       2.1       5.4
  2.     1949         23      15.3       7.8         3       4.8
  3.     1959       26.6      17.7       9.5       3.4       6.1
  4.     1969       30.1      17.7      12.4       4.5       7.9
  5.     1979       30.6      17.6      13.1         5       8.1
  6.     1985       35.1      22.2      12.9       5.2       7.7
```

**Figure 2.2** *StataQuest for DOS Editor*

### Moving From Cell to Cell

To move from cell to cell, use the arrow keys. The spreadsheet screen should be viewed as a small window through which we are looking down onto a larger spreadsheet. We can portray about 16 observations and 6 variables at a time on the screen.

Besides the arrow keys, the following keys help you move around the spreadsheet:

- The HOME key will move you back to cell A1, the top-left cell.
- The END key moves you to the bottom of the current column on the screen.
- The PGUP and PGDN keys move you one screen up and one screen down respectively.
- The ESC and F10 keys move you back and forth from the spreadsheet to the main menu (ESC) and the spreadsheet menu (F10).

### How to Input New Data into a File in the Spreadsheet

Inputting new data through the DOS version of StataQuest is slightly different from the Windows version. To input a file in DOS,

- From the *Edit* menu, choose *Spreadsheet*.
- From the *File* menu, choose *New* (SQ may choose this option for you).
- At that point SQ prompts you for the size of the file you wish to input, and you tell SQ the number of rows or observations and the number of columns, or variables.
- SQ then opens an empty spreadsheet, with missing data in each place where data are to be entered.

The next step is to begin inputting data. Using the data in Table 2.1, we begin in the top-left cell, cell x1[1]. Here we enter 1929 and press ENTER. The cursor moves automatically to cell x1[2], where we enter 1949. At this stage the basic technique is to enter the number and press the ENTER to go to the next cell down. Notice that

- We enter each number without the percent sign.
- We enter numbers without commas or dollar signs ($). These can be placed in the data by the computer as output formats (formatting information that is used when a number is output to the screen or printer). For example, we enter 6123456, **not** 6,123,456 or $6,123,456.
- Regardless of whether we enter a number like 3.0 with the decimal point and trailing zero, SQ writes the number into the spreadsheet as 3 without the decimal point and trailing zero.

When we are done, SQ will be holding the data in memory. If there is a sudden power outage, our work will be lost. Using the F10 key, we go to the SQ spreadsheet menu and from there we go to the *File* menu and *Save* the data under the name govtspen. SQ in turn saves a file on disk called *govtspen.dta*.

### Saving the File

To save a file while working in the spreadsheet, select *Save* from the *File* menu within the spreadsheet. SQ will give you a chance to supply a new data set name if desired.

## Inputting Your Data Through External (Text) Files

This command allows StataQuest to read a data file created by a non-SQ program, such as a word processor or database manager. ASCII stands for American Standard Code for Information Interchange, and ASCII files are plain character text files. Many software programs can read and write them. That allows you to transfer data from one program to another without having to type the data all over again or purchasing a conversion program.

SQ has certain specifications for the ASCII files it reads:

■ SQ assumes there is only one observation per line. Each observation can contain numeric and/or string (alphanumeric) variables. The first line can contain variable names.

■ SQ expects that files to be imported have each data element separated from the others ("delimited") by one TAB or comma. If the first line has variable names, the names should be separated by the same delimiter.

■ SQ will assume that the ASCII file has the extension .txt (ex: mydata.txt), meaning tab delimited; *.dat, meaning tab delimited; or *.csv, meaning comma delimited.

Your StataQuest disk comes with files that can be used for practice. To import a file:

■ From the *File* menu, choose *Import ASCII*.

SQ treats data that are numbers and data that are alphanumeric or non-numeric differently. Numbers can be manipulated through statistical routines. Alphanumeric data are used for labeling graphs and otherwise identifying observations ("Boston," "New York," "Pittsburgh," "Denver") but cannot be manipulated in any numeric computations. Sometimes data that are imported from another program contain words like "Male" and "Female" that the other program used in some specific way, but which SQ cannot use in its statistical routines. If you have some variable like this, go to the *Data* menu and choose *Label* and then *Strings to Labels* to convert the variable to numbers. See the section on pages 31–32 and 48. See exercises 2.3 and 2.4 at the end of this chapter.

**Note:** If coded strings like "male" and "female" are inputted, you will have to use the *String to labeled* command from the *Data* menu to convert the variable to numeric for SQ's statistical procedures to use the variable.

# Creating Random Variables

StataQuest allows you to generate random numbers easily.

## Windows

In the Windows version of StataQuest, you can generate variables to add to an existing data set, or you can generate a new data set for a simulation. To do so, choose *Generate/Replace* and then *Random numbers* from the *Data* menu. At that point, SQ presents a dialog box in which you specify the following:

■ The variable prefix, meaning that if you are generating more than one random variable based on the same distribution and if you use the prefix "ran", SQ will call the variables ran1, ran2, ran3, and so forth.

■ The number of variables to generate. Fill in a number from 1 to 25.

■ Your choice of distribution, chosen from Normal, Bernoulli, Binomial, Chi-squared, Exponential, F, Student's-t, Poisson, Uniform, Integers.

At that point, a second pop-up dialog box will appear that will ask you the appropriate information on each distribution. These are as follows:

■ **Normal.** For normally distributed data, StataQuest will ask us for the mean and standard deviation (SD) of the data. (If you are experimenting, leave those values on the default of 0 and 1.) The SD must be greater than zero.

■ **Bernoulli.** The Bernoulli distribution produces a distribution of random Bernoulli integers (0 and 1), with the 1s being distributed in a probability you specify.

■ **Binomial.** The binomial distribution is the distribution of the number of successes in N trials. If you told StataQuest already that you wanted 100 trials, now you must tell StataQuest the probability of a success. If you enter 100 trials and a probability of success (p) of 0.1, then you will have a distribution centered around the mean of 0.1 of 100 = 10. p must be between 0 and 1.

■ **Chi-squared.** The chi-squared distribution is different for each degree of freedom, so StataQuest asks you first for the number of degrees of freedom and then picks random numbers from the appropriate distribution.

■ **Exponential.** The exponential distribution is the distribution "of the amount of time until the first occurrence of an event, or, equivalently, the distribution of the amount of time between occurrences of an event, where occurrences are governed by a Poisson process" (Hays and Winkler, 1971, 230). You need to specify the mean of the distribution, which must be greater than zero.

■ **F.** The F distribution is the ratio of two independently distributed chi-squared variables, each divided by its degrees of freedom. For StataQuest to generate an F distribution, you need to specify the degrees of freedom for the numerator and the degrees of freedom for the denominator.

■ **Student's-t.** The t-distribution also varies according to its number of degrees of freedom (remember that the t-distribution is used to estimate the normal distribution when the variance is unknown and must be estimated). You must input the number of degrees of freedom.

■ **Uniform.** Uniform automatically produces uniformly distributed, pseudo-random numbers between a lower limit and an upper limit, both of which you specify.

- **Poisson.** The Poisson distribution is the distribution of a random variable, the number of occurrences in a given time period. The Poisson process has a single parameter that represents the expected number of occurrences in a single time period (Hays and Winkler, 1971, 206). You specify the mean of the distribution, meaning the number of occurrences, which is greater than zero.
- **Integer.** Integer generates random uniform integers between a lower limit and an upper limit, both of which you specify. For example, if you wanted to simulate the result of generating one die, you would enter from 1 to 6. Integer is useful for creating a simulation of a die or other games and for randomly dividing your data into arbitrary groups.

## DOS

For the DOS version of SQ, the process is almost the same:

**1** Begin with an empty spreadsheet. From the *Edit* menu, choose *Spreadsheet*.

**2** From the *File* menu, choose *New*.

**3** StataQuest asks the number of observations (rows) and the number of variables (columns). We decide to generate 100 random observations in each of 5 variables using several of the functions available in StataQuest.

At this point StataQuest provides an empty spreadsheet, 100 observations by 5 variables. The variables are labeled x1, x2, x3, x4, x5. Each cell is marked for missing data with a period ( . ).

We decide that x1 should be a uniformly random function between 0 and 1:

**1** From the spreadsheet, select the spreadsheet menu by pressing the F10 key, and then from the *Replace* menu, choose *Random numbers*.

**2** From the Random numbers submenu, choose Uniform. Uniform automatically produces uniformly distributed, pseudo-random numbers between 0 and 1.

We then continue to choose other random functions for the other variables. These work almost identically to the Windows version discussed previously and are therefore not duplicated here.

## Stata Files

SQ can read Stata files as long as they meet the file specifications of 4,000 total cells, no more than 25 variables (and 160 observations) **or** 600 observations (and 6 variables). SQ 4 saves files in Stata's version 4 format. Version 4 is incompatible with Stata's earlier formats, which range up to 3.1. If you use SQ 4 to save a file to be read by Stata or SQ 3.1 or earlier, then use the following command in the Command window (DOS: command mode): `save <filename>, old.`

# Spreadsheet and Statistical Package Files

SQ can read files from spreadsheet and statistical package programs as long as they have been converted into Stata's ".dta" format and they are within the SQ maximums of 4,000 cells, with a maximum of either 25 variables (and 160 observations) or 600 observations (and 6 variables).

The two leading conversion programs are *Stat/Transfer* and *dbms/copy*.

Spreadsheet files must have a certain form before they can be successfully converted. *Stat/Transfer*, for example, expects the following:

- A spreadsheet file will to have variable names on the first row. Variable names should be eight characters or less, begin with a letter, and include only letters and numbers, no special symbols. They should also not have any spaces.

- The second row of the spreadsheet should be the first observation of data, and it should be a complete observation, that is, no missing data or empty cells.

- Stat/Transfer assumes that whatever data is on the second row of the spreadsheet predicts the kind of data (letters, integers, numbers, and so on) in the rest of each column.

Statistical package files are normally easily converted. SPSS and SAS files, however, must be in the "portable" or "transfer" format.

# Labeling Your Data

## Windows 95

Labeling in the Windows 95 version of SQ can be done entirely from the main menu by choosing *Label* from the *Data* menu. In addition, double-clicking on any column within the Editor allows you to specify a new variable label or to rename a variable (see "Labeling from the Editor" following).

### Labeling the Data Set

SQ will allow you to input a label for the data set as a whole as long as 31 characters in length. From the main menu, choose *Label* and then *Label dataset* from the *Data* menu.

### Labeling Variables

From the main menu, choose *Label* and then *Label variable* from the *Data* menu. The label can be as long as 31 characters.

### Labeling from the Editor

Within the Editor, double-clicking (two quick clicks on a given cell) on any column of data opens a dialog box where you can (1) rename the variable; (2) input a label of as many as 31 characters (SQ does not check when the 31 character limit is reached); and (3) change the output or display format for a variable from the default %8.0g. This format means that SQ outputs as many places to the right of the decimal point as it has room for and also controls how many places to the right of the decimal point are outputted when means, standard deviations, and other statistics are computed. If you prefer that every variable and display of that variable have a consistent two places to the right of the decimal point, change the "Format:" to %9.2f.

### Labeling Each Value

StataQuest for Windows will allow you to input value labels by choosing *Label* and then *Label values* from the *Data* menu. This choice opens the first of two dialog boxes.

- **Dialog Box 1:** The Choose Variable box. In this box you decide which variable you are going to attach a label to, click on it with the mouse, and then click on OK. That action opens the second dialog box.
- **Dialog Box 2:** The Label Values box. In this box, you associate a label of eight or fewer characters with each value.
  - In the box below **Define**, we have a value listed. You choose the values through the four boxes below: First, Prev(ious), Next, and Last. These allow you to "scroll" through the values and labels one by one.
  - In the box below the words **Label (max 8 characters)**, you type a label for each value. If you type more than eight characters, SQ simply prints only the first eight.

**Note:** SQ does not print out the value once you associate a label with it. Consequently, we make it a practice to include the value inside the label. For example, if the variable *gender* is coded 1 for Males and 2 for Females, our labels might be "1-Male" and "2-Female." (The dashes [ - ] are counted as one of the eight characters.) That way we not only know the "Male" and "Female" part; we also know the values 1 and 2.

**Another example:** For a 5-value variable for education, we had labels of "1<8yrs," "2-8-11yr," "3-HSGrad" "4-SmColl," and "5-BA+." Sometimes there is room for the dash, and sometimes you have to be creative to include sufficient information in the eight characters.

**You can see the labels most readily** either in the Editor or by running a Frequency distribution by choosing *Tables* and then *One-way (frequency)* from the *Summaries* menu.

■ **TIP 2** ■ ■ ■ ■ ■ ■ ■ ■ ■ ■ ■ ■ ■ ■ ■

## Valid Variable Names

Valid variable names must be eight characters or fewer and begin with a letter. An underscore ( _ ) can be included as part of a variable name.

■ SQ's variable names are case-sensitive, that is, to SQ, *income* and *Income* are two different variables because of the capitalization of the letter I. In general, you will find it easier to work with SQ if you use all lowercase variable names.

■ SQ's developers reserve for their own use the following words, which cannot be used for variable names: _all, _b, byte, _coef, _cons, double, float, if, in, int, long, _n, _N, _pi, _pred, _rc, _se, _skip, using, with.

■ The best variable names are those that somehow convey the meaning of the variable they stand for; *income* is a much better variable name than *x10*.

## Renaming Variables

### Windows 95

Within the Editor, double-clicking (two quick clicks with the left-button of the mouse) on a cell on any column of data opens a pop-up window where the first item is the variable name for that column. You can simply input a new name at that point.

## Sorting

### Windows

Use the **Sort** button in the **Editor** to sort (ascending) on the current (where the cursor is located) variable.

**Alternative:** The Command window allows any number of variables to be used in sorting. Issue the command sort and list the variables. Example: you want to sort a survey dataset first by *gender* and within categories of *gender*, by *age*. The variables in question are called *gender* and *age*. You issue the command: sort gender age in the Command window.

## Exiting the Editor

### Windows

You can exit from the Editor by either (1) double-clicking on the system menu box in the upper left, or (2) choosing the Exit button. If you have changed an existing data set, SQ will respond with the Exit Editor dialog box, which gives you two choices:

- OK to accept the changes you made in the Editor (the changes will stay in the data in memory; you will have to save them to disk to make them permanent).

- Cancel to "restore last preserve." Restoring "the last preserve" means that the changes you made in the data since the last time you pressed the Preserve button in the spreadsheet will not be incorporated into the data in memory. If you have not pressed the Preserve button, none of the changes made during your editing session will be incorporated into the data in memory.

## ◼ EXERCISES

**2.1** Input the data set in Table 2.1. Once you are done, check each cell for accuracy and save it under the name *adc2p1* by exiting the Editor and selecting *Save* from the *File* menu.

**2.2** Input the following data set on the nations that participated in the Gulf War, excerpted from the U.N. Human Development Report:

| country | life | gdpcap | literacy | milgnp | ratmilhe | armsim |
|---------|------|--------|----------|--------|----------|--------|
| Iraq | 65 | 2400 | 89.00 | 32 | 711 | 5600 |
| Jordan | 67 | 3161 | 75.00 | 13.8 | 197 | 320 |
| Iran | 66 | 3300 | 51.00 | 20 | 377 | 1500 |
| Saudi Arabia | 64 | 8320 | 55.00 | 22.7 | 155 | 3800 |
| Israel | 76 | 9182 | 95.00 | 19.2 | 204 | 1600 |
| Kuwait | 73 | 13843 | 70.00 | 5.8 | 77 | 150 |
| U.S. | 76 | 17615 | 95.00 | 6.7 | 62.5 | 62.3 |

The variables are as follows: *life*, life expectancy in years at birth, 1987; *gdpcap*, gross domestic product per capita 1987; *literacy*, percent of the adult population estimated as able to read and write 1985; *milgnp*, military spending as percent of Gross National Product; *ratmilhe*, ratio of military spending to health and welfare spending in $; *armsimp*, arms imports in 1987 in U.S. $ millions. Save the file under the name *adc2p2*.

**2.3** Use the *Import ASCII* command from the *File* menu to import the file *import1.asc*. This file is included with your StataQuest disk.

**2.4** Add a data set label and variable labels to the file saved in problem 2.1, as per the description of the variables above. Are value labels needed? Why or why not? Press F3 or choose *Describe variables* from the *Summaries* menu to see the labels you are inserting. Rename the variables to descriptive names of your choice if you did not do so in problem 2.1. Save the revised file as *adc2p4*.

**2.5** Add a data set label and variable labels to the file saved in problem 2.2, as per the description of the variables above. Are value labels needed? Why or why not? Press F3 or choose *Describe variables* from the *Summaries* menu to see the labels you are inserting. Save the revised file as *adc2p5*.

# 3

## Modifying Your Data

## Why Change Data?

You will almost always have some change you want to make to your data. You may want to change the data so that it will fit within a certain table format. You may want to have a dichotomous variable coded 0 and 1, but your present variable is coded 1 and 2. You may want to have a variable that is lagged one time period. Or the data may be coded in a five-point scale from strongly dislike (1) to strongly like (5), but there are also codes of 8 (no opinion) and 9 (question was not asked by the interviewer), and you want to ensure that the 8s and 9s are coded for missing data so that they will not be counted in calculating statistics in a table. Many reasons for changing the data ensure that the codes for the data match the assumptions of the particular statistical test. Thus, changing the data in ways that improve the validity of statistical tests is a standard activity in data analysis. StataQuest consequently provides many ways to change your data.

In all versions of StataQuest, you use the Editor for making some changes and you make others directly without entering the Editor. In each section below, we will describe how to make these changes according to the version of StataQuest you are running. Here are the typical activities you might do in this chapter:

■ ■ ■ ■ ■ ■ ■

# IN THIS CHAPTER

■ **Changing cell contents**

■ **Adding variables or observations**

■ **Deleting variables or observations**

■ **Formulas and functions**

■ **Recoding data**

■ **Subsetting data**

■ You find one or a small number of mistakes. You decide to correct them within the Editor. This is a good idea with small numbers of corrections.

■ You want to add a variable.

■ You obtain some additional observations and want to add them to an existing data set.

■ You want to add a new variable that is some function of an existing variable.

■ You want to delete a variable.

■ You want to delete an observation.

■ You want to delete all the observations that meet some criterion.

■ You have a variable such as education that is coded as the number of years of education for each respondent. You want to recode those years into a variable where 1=8 years or less; 2=some high school; 3=high school graduate; 4=some college; 5=college graduate.

■ You want to do some analysis on a portion of the data—a subset—meeting some criterion.

# Changing Cell Contents

Changing a cell's contents is typically something you do when you have made a mistake entering the data, and you want to correct one or a few cells, or the data set is very small, and you are making changes to an entire variable or observation.

## Changing One Cell

You find that there is an error in your data. You made a mistake in a cell when you typed in a data set into StataQuest. To fix it, you enter the Editor, go to the appropriate cell, make the change, exit the Editor, save the revised file if the change is one you want to have the next time you run StataQuest, and go on to your next analysis.

### *Windows*

**1** If the file is not already in memory, read the file into memory by choosing *Open* from the *File* menu.

**2** Click on the **Editor** button or choose *Edit Data* from the *Data* menu to enter the Editor.

**3** Use arrow keys or scrolling bars to go to the appropriate place in the spreadsheet.

**4** Type the new value.

**5** Exit the spreadsheet by clicking on the **Exit** button.

**6** Tell SQ in the Exit dialog box that you want to keep the edited or changed values.

**7** Save the file by choosing *Save* or *Save As* from the *File* menu.

### *DOS*

**1** From the *Edit* menu, select *Spreadsheet* to enter the Editor.

**2** Move the cursor to the appropriate cell with the arrow keys.

**3** Type the new value.

**4** Exit the Editor with ESC or by choosing *Quit editor* from the *File* menu.

**5** Save the file by choosing *Save* from the *File* menu.

**Remember that when you make a change through the Editor, the change is made to the data you have in memory, not to the data you have saved on disk.**

## Changing Groups of Cells

You can change groups of cells in the same way that we changed a single cell above, by using the Editor to make the change. There are circumstances under which this can be dangerous, however. If your keyboarding skills are poor or you inadvertently hit some other key on the keyboard, you might change some other cell by mistake. Good data analysts take care to minimize the probability that these kinds of things can happen. One way to do this is to use the Editor to change only the observations that need to be changed, not every observation. The Windows version of StataQuest provides a way to edit just the observations you are interested in.

Suppose, for example, you find that all of your observations that are male (variable *gender*, code 1) have an error in either the variable *gender* or some other variable. You want to correct that error in the Editor.

### Windows
You can look at only those observations that are *male* by using the Command window to ask SQ to edit only a portion of the data. Typing the word edit in the Command window is equivalent to pressing the Editor button or choosing *Edit data* from the *Data* menu. Here we edit only the observations where the variable *gender* is equal to 1. **Note that the double equals sign is correct.**

---

### Stata Command

```
edit if gender == 1
```

---

**You can amend the command "edit" either with "if" statements or with "in" statements.**

**"if" statements:**
- "if" statements select observations on the basis of one or more variables.
- "if" statements always take the form of **if <variable> <operator> <value or variable>**, where
  - <variable> is a variable name
  - <operator> is one of the following:
    - equals (==)
    - not equals (~=)
    - greater than (>)
    - less than (<)
    - greater than or equal to (>=)
    - less than or equal to (<=)
  - <value or variable> is either a value such as "1" or "2" or another variable name.

Ex #1:
```
if gender == 1
if educat > 12
if educ1 == educ
```

Ex #2: You have a variable *educ* in your data set that is coded as the number of years of education. You want to edit only those people with some college or a Bachelor's degree. In the Command Window, you write:

```
edit if educ > 12        or
edit if educ >= 13        [Both have the same effect]
```

- "if" statements can be connected with "and" (&) and "or" ( | ):

Ex: You have a data set with a variable for gender called "gender" and a variable for education called "educ." These are coded as the previous example. You want to make some changes in the Editor for males who have less than a high school diploma. You write

```
edit if gender == 1 & educ < 12
```

**"in" statements:**

- ▪ "in" statements select observations for editing in the Editor on the basis of the observation number as the data set is currently sorted. (Be careful: some SQ procedures sort the data and leave it in a different order from what you may recollect.)

- ▪ "in" statements take the form of **in <value> <[ / value ]>**, where

  <value> is an observation number

  < / value> means that the first observation number can be followed, if desired, by a second observation number separated by a slash.

  Example: 1/10, meaning observation numbers one through ten.

- ▪ Ex: `edit in 1`                              edits the first observation ("obs")
  `edit in 10`                           edits the 10th obs
  `edit in 10/35`                      edits the 10th through 35th obs
  `edit in 1/1` (that is, one/el)   edits the first through last obs
  `edit`                                     same
  `edit in -2`                           edits the second to the last obs
  `edit in 10 / -3`                   edits the 10th through second to last obs

- ▪ "in" statements may be combined with "if" statements:
  Ex: `edit if gender ==`          edits if gender equals 1 in the 10th
  `1 in 10/22`                        thru the 22nd observation

### DOS

The DOS version of StataQuest does not allow you to edit a portion of the data. When you enter the spreadsheet by choosing *Spreadsheet* from the *Edit* menu, the entire data set is available.

Other StataQuest commands entered through command mode, however, do allow the "if" and "in" statements explained earlier.

# Adding Variables or Observations

The easiest way to add variables in all versions of StataQuest is to use the Editor. Enter the Editor by clicking on the Editor button (Windows) or by selecting *Spreadsheet* from the *Edit* menu (DOS).

## Windows

**To add a variable:**

**1**   Go to the top of an empty column.

**2**   Start typing.

---

## ■ TIP 3 ■ ■ ■ ■ ■ ■ ■ ■ ■ ■ ■ ■ ■ ■

### How to Use Command Mode

Writing a command in StataQuest's Command Window is straightforward, but there are several rules to remember.

■ First, all commands are in lower case. StataQuest is **case-sensitive**. The command "describe" is different than "Describe" is different than "DESCRIBE." The same is true of variable names: x is different than X.

■ Commands are **not** entered with quotes around them.

■ Commands have several parts: the command itself, a list of one or more variables, any "if" or "in" statements, then a comma, and then one or more options if required. Here is a graph command with three options:

```
graph educ income if gender == 1 , xlabel ylabel ti
("Sample Graph")
```

■ Second, in general, command mode has very **flexible spacing** rules. You can leave spaces where you want.

■ Third, you can **recall any previous command** in command mode with the **PREVLINE** and **NEXTLINE** keys. These are **PGUP** and **PGDOWN** in Windows and DOS. Once the command is on the command line, you can edit it and press ENTER to execute it again. The following keys work when you are editing a command: all of the **arrow keys, HOME, END, INS, DEL**.

---

**3** Use ENTER to finish each cell entry. When you finish the current column, SQ will remember how many observations you have in the data set and go to the beginning of the next column.

**To add one or more observations:**

**1** Go to an empty row.

**2** Start typing.

**3** Use TAB to finish each cell entry. When you finish the current row, SQ will remember how many observations you have in the data set and go to the beginning of the next row.

## DOS

**To add a variable:**

**1** Decide where you would like the new variable. Place the cursor on the variable to the right.

**2**  Press F10 to activate the *Spreadsheet* menu. From the *Add* menu, choose *Variable*. Enter a new variable name. Press ENTER. SQ then adds the appropriate column with missing data, placing the cursor at the top.

**3**  Enter data cell by cell, concluding each cell with the ENTER key.

**To add one or more observations:**

**1**  Decide where you would like the observation to be.

**2**  Use F10 to activate the *Spreadsheet* menu. From the *Add* menu, choose *Observation*. Press ENTER. SQ then adds the appropriate row where your cursor is located.

**3**  Enter data cell by cell, concluding each cell with the right arrow key.

**Other DOS options.** When you choose *Add* from the *Spreadsheet* menu, you have four options:

- **Variable.** To add a variable.
- **Observation.** To add a row or observation in the current location.
- **Obs at end of data.** To add a row or observation at the end or bottom of the data set in memory.
- **Cell.** To add a cell to the current variable in its current location, shifting everything in the current variable down one cell. This option is for correcting typing input errors.

# Deleting Variables or Observations

The easiest way to delete variables or observations in all versions of StataQuest is to use the Editor. Enter the Editor by clicking on the Editor button (Windows) or by selecting *Edit / Spreadsheet* (DOS). **Place the cursor in a cell in the row or column that you want to delete.**

## Windows

**1**  Enter the Editor by clicking on the Editor button.

**2**  Go to the appropriate column or row, that is, the one you want to delete, and choose Delete from the Editor button bar.

**3**  A dialog box will open, asking whether you want to:

- "delete variable var_name"—this option deletes the **variable**, the **column** of data on which the cursor is located.
- "delete observation obs#"—this option deletes the **row** of data on which the cursor is located.

- "delete all observations where the current variable is equal to its current value" —suppose the cursor is located on the variable *gender*, in an observation which is male (*gender* is equal to 1). Selecting this option would delete **all observations where gender equals 1**.

## DOS

**1** Enter the Editor by selecting *Spreadsheet* from the *Edit* menu.

**2** Use the arrow keys to place the cursor on the column or row you want to delete.

**3** Use F10 to access the *Spreadsheet* menu. From the *Drop* menu, choose one of the following:

- **Variable.** To delete the current variable.
- **Observation.** To delete the current observation.
- **Observations.** To delete all observations where the current variable is equal to its current value. Thus, if the cursor is located on the variable gender, coded 1 for males and 2 for females, in a cell which is coded 1, selecting this option would delete all observations that are coded 1 in the gender column.
- **Cell.** To delete the current cell, shifting all cells below it in the current variable or column up one. This option is for correcting typing input errors.

# Formulas and Functions

All versions of StataQuest allow you to transform variables into new and different variables with formulas and functions. Functions are short-cuts for formulas, allowing you to do something very easily that might be difficult in a formula. The function abs(var_name), for example, takes the absolute value of whatever variable is named in the parentheses.

The last chapter discussed another way to use formulas and functions, choosing to define random data through StataQuest's menu choices. In Windows, we chose *Generate/Replace* and then *Random* from the *Data* menu. In DOS, we pressed F10 to access the spreadsheet menu (from within the spreadsheet) and then chose *Random numbers* from the *Replace* menu.

An important difference between the Windows and DOS versions of SQ is that the DOS version will allow you to apply a formula to an existing variable, thus changing it perhaps irrevocably, but the Windows Formula dialog boxes insist that you define a new variable for the formula, even if the new variable is a function of the old variable.

## Windows

Choosing *Generate/Replace* and then *Formula* from the *Data* menu allows you to define a new variable and input its formula in one step. The **Generate Formula** dialog box opens; you fill in a new variable name and the appropriate formula in the two boxes.

- **New variable names** should be one to eight characters long, begin with a letter, and be composed of letters, numbers, and the underscore (_). Don't begin your variable names with an underscore ( _ ). The following variable names are reserved for SQ's own use: *byte, double, float, if, in, int, long, using, with*. See Tip #2—*Valid Variable Names*, in Chapter 2.

- **For the formula box,** you write in what is to the **right of the equals sign** in your formula. Examples:

| If the formula is: | You write in the box: |
| --- | --- |
| poplog1 = log(pop90) | log(pop90) |
| educ1 = educ | educ |
| lagtime1 = time1[_n-1] | time1[_n-1] |

Here is a dialog box example of adding a variable to the data set and defining it using a formula. We use the data set called *urbschls*, which is provided with StataQuest. This data set contains variables for the overall and minority percentages of both the city as a whole and the school system for 26 large cities in the United States. We will add to the data set a variable for the **minority population percentage** in each city. This variable is composed of the African-American population + the Latino population + the other minority population, with the total then divided by the total population of the city, and the resulting decimal is multiplied by 100. The formula will be 100 * ((popaa + pophisp + popother) / poptot), where *popaa, pophisp, popother,* and *poptot* are variable names.

Note that we have used parentheses liberally in developing the formula, similarly to what one would do in algebra. Spacing between each variable name or parenthesis is up to the user. StataQuest is flexible enough to take spaces between words or not, depending on your preference. In the formula, we have left a space next to each symbol or variable name to clarify the presentation.

**1**  From the *Data* menu, we choose *Generate/Replace*.

**2**  We then choose *Formula*.

**3**  The result is the following dialog box, which we fill in as shown in Figure 3.1.

**4**  Type the new variable name: minper.

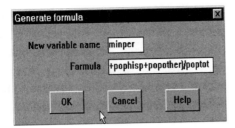

**Figure 3.1**  *The Generate Formula Dialog Box*

**5**   Type the formula: `100*((popaa+pophisp+popother)/poptot)`. Notice that the formula scrolls off to the left if it is too long for the dialog box.

**6**   Click on "OK" to complete the action.

We can check our results by using the Editor to look at the variable.

## Working with Formulas

SQ will allow you to redefine a variable using standard algebraic formulas.

**Arithmetic.**  All the standard tools of addition, subtraction, multiplication, and division are present:

*Addition*          +
*Subtraction*       –
*Multiplication*    *
*Division*          /

**Exponentiation.**  You can also raise a number to a power. If you raise it to the power of 2, you are squaring it; power of 3, cubing it; power of 1/2, taking its square root; power of 1/3, taking its cube root.

*Exponentiation*   ^

If var1 is a variable name, and we are going to transform it into var2 with one of the preceding powers, we could, for example:

*Square it*               var2 = var1 ^ 2
*Cube it*                 var2 = var1 ^ 3
*Take its square root*    var2 = var1 ^ .5
*Take its cube root*      var2 = var1 ^ .33

**And, Or, Not.**  In addition, we can use three *logical operators* in our formulas:

*And*    &
*Or*     |
*Not*    ~

**Some useful formulas.** We can combine the addition, subtraction, multiplication, and division signs with the logical operators to make some useful formulas:

- **Change the scale on a variable.** We have data on Gross National Product per capita to the exact dollar. We need the data only to the nearest thousand dollars. Our new variable is *gnpnew*, and the formula to the right of the equal sign is `gnpold / 1000`.

- **Compute an average.** We have data on the number of federal grants received by each county for the last several years. We want the average number received each year. Our new variable will be called *avggrant*, and the formula to the right of the equal sign is `(no89 + no90 + no91 + no92) / 4`, where *no89* is the number of grants received in 1989, and so on.

- **Lag a variable.** We have time series data on the Gross National Product in the United States since 1930. We want to create a variable that is lagged one year so we can do a time series analysis. Our variable is called *gnp*. We will call the new variable *gnplag1*. The formula to the right of the equal sign is `gnp[_n-1]`. To lag a variable twice, we would write `_n-2`, and so on.

**If a formula is too long for the dialog box.** As you type, the formula will scroll to the left.

**Writing formulas in the Command window.** Write the entire formula, including the variable name and the equals sign, preceded by the word `generate`, meaning that you are going to create a new variable:

---

**Stata Command**

```
generate avggrant = (no89 + no90 + no91 + no92) / 4
```

---

**Functions.** StataQuest has many functions that can be placed in formulas. They are listed in Appendix A.

## DOS

To enter a formula in the DOS version of StataQuest, you must be in the spreadsheet. Follow these steps:

**1** Enter the spreadsheet: from the *Edit* menu, choose *Spreadsheet*.

**2** Add a new variable if necessary by pressing F10 to access the spreadsheet menu, and then choosing *Variable* from the *Add* menu.

**3**  Position the cursor in the column of the variable that is to have the formula applied to it.

**4**  From the *Replace* menu, choose *Formula*.

**5**  Enter the formula, remembering to enter only the portion to the right of the equals sign. Press ENTER when finished.

All the functions used in the Windows edition of SQ work the same way in the DOS version.

To enter a command similar to the Windows Command window, you must exit menu mode by choosing *Command mode* from the *File* menu. You then enter commands as if you were in the Command window. See Tip 3 on command mode on page 41.

## String to Labeled

The *string to labeled* option is used to transform data that have been input as names (alphanumeric or string data) to numeric data. If you had a variable for gender where the codes entered were *Male* and *Female*, this option enables you to convert these codes to 1 and 2 respectively, with Male and Female being retained as labels for these numbers. You would want to do this if you had some string variable that you wanted to use in a statistical procedure; StataQuest's statistical procedures will not operate on string variables, and you would receive an error message.

**Note:** The "String to labeled" dialog box will not list any variables unless one or more string variables are in the data set. When a variable is transformed from string to labeled, the labeled variable takes the old string variable name, and the string variable is deleted from the data set.

## Labeled to String

The *Labeled to string* option is used to transform numeric data that has value labels attached to the numbers to string data. This option is generally rarely used, but operates similarly to *String to labeled*.

# Recoding Data

When you recode data, you change the codes from those you don't want to those you need for some purpose. A major purpose is to simplify a variable with detailed coding (like the number of years of education) into a reduced number of categories. Here is an example of an education variable being transformed in this manner.

---

## ■ TIP 4 ■ ■ ■ ■ ■ ■ ■ ■ ■ ■ ■ ■ ■ ■ ■

### When Is a Number *Not* a Number?

Sometimes students enter numbers, but take some other action that makes SQ and spreadsheet software treat those numbers as letters, or an alphanumeric or string variable. How would you know that this is the case? The easiest way to tell is if any missing values are blank (indicating an alphanumeric) rather than a period ( . ), indicating a numeric variable. Or you may find that SQ's procedures are giving you error messages that there are "no observations."

   If you have data that are numbers, but were entered inadvertently as strings, do the following:

**1**   Choose *Generate/Replace* and then *Formula* from the *Data* menu.

**2**   Input a new variable name.

**3**   Input the following formula: `real(string-var-name)`, where the name inside the parentheses is the name of the string or alphanumeric variable you want to transform into a numeric variable.

**Warning:** the *real()* function works only with numbers that were inadvertently entered as strings.

---

StataQuest has a Recode dialog box to help you do recodes. To do a recode:

- First, ensure that you have the space in your data set to add a new variable. Recode requires that you add a new variable to your data set. Choose *Describe variables* from the *Summaries* menu to see how close to the 4,000 data point capacity of StataQuest you are, and if necessary drop a variable (by entering the Editor and choosing *Delete*) to make space.

- Second, have some idea of how you want to recode the existing codes for the variable. You can obtain information on the existing codes for a **categorical** variable by choosing *Tables* and then *One-way (frequency)* from the *Summaries* menu to run a frequency distribution for the existing variable. You can get information on a **continuous** or **measurement** variable by choosing *Median/Percentiles* from the *Summaries* menu. The latter will give you cutoffs for the 1st, 5th, 10th, 25th, 50th, 75th, 90th, 95th, and 99th percentiles. Sometimes, these cutoffs can be used directly as the cutoffs on a recode.

- Third, once you know how you want to recode the variable, begin the recode procedure by choosing *Generate/Replace* and then *Recode* from the *Data* menu. This opens the **Choose Variable** dialog box, the first of two. In this box, you specify three things:

- **The variable you wish to recode.** You specify this by selecting a variable from the list provided under the words "Recode values of".
- SQ asks "into ____ values (or classes)." This number is the the **number of values** you wish the resulting variable to have.
- "**New variable name: _____**" SQ will define a new variable with the number of values or classes you specified in question #2.

■ Click on "OK" (Windows) or complete the questions (DOS).

■ Fourth (Windows), the **Recode Values** dialog box will now appear. In it, SQ tells you the variable you are recoding and asks you to specify the values being recoded ("Value Range") and the value they are being recoded into ("Recoded Value"). If we want to recode the age variable into categories including a category for those 18 to 35, and that category should have the value 1, then we would enter 18 and 35 into the Value Range, and 1 into the Recoded Value, as illustrated in Figure 3.2.

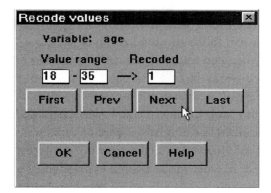

**Figure 3.2** *Recode Values Dialog Box*

■ You can use "max" to indicate the maximum value, "min" to indicate the minimum value, and "*" to indicate all other values to simplify recoding.

■ Complete the Recode Values dialog box by proceeding through the new codes one by one, choosing Next each time and using Prev (for "Previous"), First, or Last to ensure that the codes are entered accurately.

■ Choose OK when the Recode Values dialog box is finished. SQ will then generate the new variable.

■ (DOS) Choose Recode from the Replace menu within the spreadsheet. Process with the steps listed for Windows. You must start with the cursor in the column of the variable you want to recode.

### Recoding Data in Command Mode

The code for this command is the word "recode," followed by the variable name, and then the recodes. Here we recode gender from codes of 1 and 2 to codes of 0 and 1 by recoding the code 2 to the code 0: recode gender 2 = 0

SQ understands the recode command through the placement of the equal sign (=). To the left of the equal sign are the old codes; to the right, a single new code. Spacing is flexible. Next we recode an education variable where the old codes are the number of years of education; the new codes are to be 1 for 8th grade or less; 2 for "some high school"; 3 for high school graduate; 4 for "some college"; 5 for college grad or more:

```
recode educ  1/8=1  9/11=2  12=3  13/15=4   16/ max = 5
```

- You can abbreviate a series of numbers with the first and last number connected by a slash ( / ).

- You can use the words "max" or "min" to signify the maximum or minimum number in the series, as you can in the dialog boxes earlier.

- You can recode only one variable at a time.

- You can use the abbreviation * to indicate all other codes not referenced already in your command (see the second example).

Here are some examples using an imaginary variable called *x1*.

| | |
|---|---|
| recode x1 2=1 4=0 8 9 = . | Note that 8 and 9 are changed to missing. |
| recode x1 2=1 4=0 * = . | After changing to 1 and 0, all others are changed to missing. |
| recode x1 5/25 = 1 26/35 = . | 26 through 35 are changed to missing. |
| recode x1 5 7 9 12 22 25 = 0 | all changed to zero. |
| recode x1 5 10 20/25 = 0 | 5 and 10, plus 20 through 25, changed to zero. |
| recode x1 min/10 = . | Lowest number through 10 changed to missing. |
| recode x1 . = 8 | Missing values changed to 8. |
| recode x1 7 8 9 = . | 7, 8 and 9 changed to missing. |

## Subsetting Data

Data analysts often want to analyze subsets of the data separately. You may want to perform one analysis on the men and another on the women. You may want to perform one analysis on Southern states and another on the balance of the nation. You may want to perform one analysis on the OECD nations and another on the third world.

Within StataQuest, there are two ways to analyze subsets of data:

- **The first method is to use the Editor** to drop all observations not part of the subset. Perform the analysis on the subset, the data that remains. Save the subset as a new data set with a different name. If you need to analyze other subsets of the data, repeat the process.

- **The second method is to use command mode** to recall the previous command with the PREVLINE key (PGUP in Windows and DOS). Add an appropriate "if" or "in" statement before the comma. Rerun the command on the subset indicated with the "if" or "in" statement by pressing ENTER. Repeat as necessary.

If you need to do only one or two analyses of a subset, method #2 may be easier. If your analysis will be extensive, it is probably easier to save subset files with method #1.

## Method #1: Forming Subset Files

Here is a demonstration of Method #1. You are using *hrsamp94*, which is a data set composed of 108 randomly selected members of the U.S. House of Representatives in 1994. You are interested in the relationships between campaign spending in the district (*expwin*) and the vote for President Bush in the congressional district in 1992 (*bush92*). You decide to run separate graphs for the Republicans and the Democrats using the *party* variable, coded 1 for Democrats and 2 for Republicans.

**1** Enter the Editor by clicking on the Editor button.

**2** Position the cursor in the *party* column, on one of the cells that is coded 1.

**3** Click on **Delete** from the Editor button bar.

**4** A dialog box will ask whether you want to:
- Delete variable *party*
- Delete observation #2
- Delete "all 46 obs. where *party* == 1"

**5** Choose the third of these, leaving a data set in memory consisting only of Republicans.

**6** Carry out whatever analyses you wish on your subset, saving the data (*File / Save As*) under a new name if you want to have the subset available for further analysis at a subsequent time.

**7** Repeat the process with *hrsamp94*, positioning the cursor on a cell that is coded 2 to produce a subset consisting of Democrats.

## Method #2: Adding "if" or "in" to the Previous Command in the Command Window

Here is a demonstration of method #2. We run a graph from the human-dev.dta data set, a scatterplot of variables *life* and *gnpcap87*, with *name* being a variable that contains a symbol used to name each point. We then press the PGUP key to retrieve the previous command in the Command window. Here it is:

---

### Stata Command

```
. graph life gnpcap87, xlab ylab bor symbol([name])
```

---

We then use the arrow keys to position the cursor just to the left of the comma, and type in the text if world == 1. We press ENTER whenever we finish altering the command.

---

### Stata Command

```
. graph life gnpcap87 if world == 1, xlab ylab bor symbol([name])
```

---

A new graph appears in the graph window, restricted only to those observations for which gender equals 1. We could then use the same technique again and change the "world == 1" to "world == 2" and "world == 3" to see the results for the other codes.

So the general technique is the following:

**1** Use the mouse to run the command on all observations.

**2** Use PGUP to retrieve the command in the Command window.

**3** Alter the command in the Command Window by inserting an appropriate "if" or "in" statement before the comma. If there is no comma, insert the "if" or "in" statement at the end of the command.

**4** Do not insert a second comma.

**5** Press ENTER to re-execute the command, this time on the subset of data.

**6** Repeat as necessary.

## DOS

The second method cannot be used in SQ's DOS environment.

## ■ EXERCISES

**3.1** The data set *gsssurv.dta* has a variable in it for the age of the respondent. Add a new variable to the data set, and recode that variable into three categories, one for those younger than 35 years of age, the second for those 35 to 54, and a third for those 55 and older. Code the three categories 1, 2, and 3, respectively, and include value labels. You will have to delete another variable first to do this because *gsssurv.dta* takes almost 3,600 of the 4,000 cells allowed in StataQuest. (Delete the income variable.) Save the new data set under the name *adc3p1*.

**3.2** Using the data set *transit.dta*, add variables to the data set for the following and recode as indicated: *trips* (2 categories, cutoff at 1 million trips); *fare* (2 categories, top category for those with fares above $1.00 per ride); *budget* (2 categories, top category above $500 million); *fedlsub* (2 categories, top category above $20 million). Your new data set should have both the old variables in it and the new one (for a list of variables, from the *Summaries* menu, choose *List variables*). Save the new file under the name *adc3p2*.

**3.3** Using the data set *humandev.dta*, use the world variable to define three subsets: one for the industrialized nations, the second for the newly industrializing nations, and a third for the third world. Save each as a separate file, giving each the names *adc3p3a*, *adc3p3b*, and *adc3p3c*.

# 4

# Summarizing and Examining Your Data

## The Basics of Data Analysis

In this chapter, we cover the two topics: (1) what is in a data set, and (2) how can we obtain the basic statistics—mean, median, percentiles—from a data set?

## What Is in a Data Set?

StataQuest provides three ways to find out what is in a data set:

- You can and should obtain a description of the data in memory, along with a list of the variables.
- Additional data set information has been specifically added to each data set provided with SQ, and you can add your own information for future reference to any SQ data set.
- You can list the data from any data set in memory.

■ ■ ■ ■ ■ ■ ■

## IN THIS CHAPTER

■ **What is in a data set?**
Describe variables
Dataset information (Windows only)
List data

■ **Basic statistics**
Means and standard deviations
Means and standard deviations by group
One-way means and standard deviation
Two-way means
Median and percentiles
Confidence intervals

## Describe Variables

To obtain the size of your data set in memory, along with a list of its variables: from the *Summaries* menu, select *Describe* variables.

This command results in the number of rows/observations in the data set, the number of columns/variables in the data set, the variable names, the variable types (alphanumeric or numeric), the output format for each variable, and the variable label, if one has been input.

Figure 4.1 shows an example of the resulting output. In the first four lines, SQ for Windows includes the following (SQ for DOS omits these four lines):

**Line 1:** "Contains data from C:\SQDATA\TRANSIT.DTA" Our data set is found on the C: drive, in the sqdata subdirectory, and the full name of the file is *transit.dta*.

**Line 2:** Our data set has 10 observations (rows); the maximum number of rows is 600. Our data set label is "Urban Mass Transit Systems."

**Line 3:** Our data set has 6 variables (columns); the maximum number of columns is 25. We last saved this data set on July 18th, 1995, at 9:10 p.m.

**Line 4:** Variables times observations is 60; SQ's maximum is 4,000.

SQ then has variable information, with one line provided per variable. This information is divided into columns:

**Column 1:** The variables are numbered, and each has a name of eight characters or less.

**Column 2:** The terms "str15" and "float" refer to the variable types. These are the following:

• **Numeric types:** byte, int, long, float (the default), and double.

```
┌─────────────────────────────────────────────────────────────────┐
│ ▓▓Stata Results▓▓▓▓▓▓▓▓▓▓▓▓▓▓▓▓▓▓▓▓▓▓▓▓▓▓▓▓▓▓▓▓▓▓▓▓▓▓▓▓▓▓▓▓▓▓▓▓▓▓ │
├─────────────────────────────────────────────────────────────────┤
│                                                                 │
│   . describe                                                    │
│                                                                 │
│  Contains data from C:\SQDATA\TRANSIT.DTA                       │
│    Obs:    10   (max=  600)        Urban Mass Transit Systems   │
│   Vars:     6   (max=   25)        18 Jul 1995 21:10            │
│    V*O:    60   (max= 4000)                                     │
│    1. system          str15 %15s   City Transit System         │
│    2. trips           float %9.0g   No. of Daily Trips         │
│    3. fare            float %8.2f   Minimum Fare Per Ride       │
│    4. budget          float %9.0g   1991 Budget in $ Millions   │
│    5. fedlsub         float %8.1f   Federal Subsidy in $ Millions│
│    6. subperc         float %8.1f   Federal % of Local Budget   │
│  Sorted by:                                                     │
│                                                                 │
└─────────────────────────────────────────────────────────────────┘
```

**Figure 4.1** *Describe Variables*

- **Non-numeric type:** string, meaning any combination of letters and numbers. Strings are denoted strxx, where xx is a number indicating the maximum width of the string.

  **Note:** beginning users can ignore these types. They become important with large data sets, when hard disk space is at a premium and efficient processing is crucial. But with data sets the size of StataQuest's, they can be ignored.

**Column 3:** The output format for a variable. These can be changed by double-clicking on any cell in the column you want to change in the Editor (Windows) or through the format command (DOS). %9.0g is the numeric default and outputs as many places to the right of the decimal point for which there is room. %9.2f allows two places to the right of the decimal point, %8.1f, one place, and so on. String variables have formats indicating the number of columns to be output; "%15s" indicates a string variable with as many as 15 columns displayed.

**Column 4:** A variable label, if one has been input, of 31 or fewer characters. Enter a variable label by double-clicking on the relevant column in the Editor (Windows) or from the spreadsheet menu item *Labels*, by choosing *Variable* (DOS).

## Dataset Information (Windows Only)

Information on the origin of every data set included with StataQuest has been included and can be listed on screen by the following: from the *Summaries* menu, choose *Dataset information*.

This information is provided by the authors and is not automatically present in every SQ data set. Figure 4.2 is the Results window after running the Dataset information command.

You can include your **own** information with each data set by writing "notes" (our data set information has been included as a "note" also) in the Command window. You can include any message you want, as the Command window following Figure 4.2 indicates.

---

## Stata Results

```
. notes

_dta:
  1. transit is a data set drawn from a small table in the New York Times,
  2. Tues., 4/16/1991, p. A12, entitled 10 Transit Systems Facing Trouble.
  3. It can be used to study the relationship between the fare charged
  4. and the transit system budget, the size of the federal subsidy,
  5. and the per cent of the local budget that the federal subsidy covers.
  6. Several simple graphs can be produced from this data set.
  7. No categorical variables. The name of the city is the ID variable.
  8. 10 Observations; 6 variables.
```

**Figure 4.2** *Notes on transit.dta*

---

## Stata Command

```
note: This data set is interesting / use for paper this quarter.
```

---

The term "note" must be in lower case; the colon ( : ) is required. After the colon, you can include as many as 255 characters, approximately the equivalent of 4 lines of information. Additional notes can be included if the 255 characters do not suffice.

If we add this note to the transit data set, our new transit data set notes would look like the Results window in Figure 4.3.

Notes can be general, attached to the data set (thus the printout of the term "_dta" in Figure 4.3) or they can be attached to individual variables. If you want to attach a note to an individual variable, you list the variable's name after the term "note" and before the colon ( : ). See the example for the variable *trips* on the next page. As an exercise, input a note on a variable of your choice:

```
  Stata Results

  . notes

  _dta:
    1. transit is a data set drawn from a small table in the New York Times,
    2. Tues., 4/16/1991, p. A12, entitled 10 Transit Systems Facing Trouble.
    3. .... [Lines 3 through 7 omitted]
    8. 10 Observations; 6 variables.
    9. this data set is interesting / use for paper this quarter.
```

**Figure 4.3** *Notes for transit.dta with User Addendum in Line 9*

**1** Use the Command window to type: note: <content of your note>
**2** To see the result, from the *Summaries* menu, choose *Dataset* info.
**3** If the note is important to have in the future, save the data set. From the *File* menu, choose *Save* or *Save As*.

```
  Stata Command

  note trips: is it really true that NYC has 3 times the no of
  trips of other systems?
```

## List Data

We can look at the data in any SQ data set in memory by selecting *List data* from the *Summaries* menu. When we choose to list the data, we have two basic formats.

▪ First, we can choose to list as many as six variables in neat columns. After six variables,* SQ uses a format that lists all of the data for each case before going on to the next case. Each variable name is listed, then each value. Figure 4.4 provides an example.

▪ We can list all the data for each case, arranged neatly, but not in columns. This is the format chosen when more than 6 variables are specified. Figure 4.5 provides an example.

---

* The cutoff of six variables assumes variables of normal "width." If the variables chosen include string variables (names) with widths wider than eight columns, fewer than six variables may be possible.

## ■ TIP 5 ■ ■ ■ ■ ■ ■ ■ ■ ■ ■ ■ ■ ■ ■ ■ ■ ■

### Alphanumeric (String) Variables in Statistical Routines

An alphanumeric/string variable cannot be used in statistical routines. If you have an alphanumeric variable that you want to use in a statistical routine (for example, a variable called gender entered as *male* or *female* that you want to use in an analysis of variance procedure), you can convert the variable as follows:

■ **Windows users:** From the *Data* menu, choose *Generate/Replace*, and then *Strings to labels*.

■ **DOS users:** Go to the spreadsheet, place the cursor on the column with the variable you want to convert, and then from the *Replace* menu, choose *String to labeled*.

The result will be a variable coded as integers, with the old string codes *male* and *female* now present as value labels.

---

### Stata Results

```
                make     price     mpg    rep78     hdroom     trunk
1.      AMC Concord      4099      22       3          2.5       11
2.        AMC Pacer      4749      17       3          3.0       11
3.        AMC Spirit     3799      22       .          3.0       12
4.     Buick Century     4816      20       3          4.5       16
5.     Buick Electra     7827      15       4          4.0       20
6.     Buick LeSabre     5788      18       3          4.0       21
```

**Figure 4.4** *"List Data" in Neat Columns (Six or Fewer Variables Specified)*

---

### Stata Results

```
Observation 10

    make  Buick Skylar..    price     4082         mpg          19
   rep78             3      hdroom     3.5        trunk          13
  weight          3400      length     200         turn          42
   displ           231      gratio    3.08      foreign    Domestic
```

**Figure 4.5** *List Data by Observation (More Than Six Variables Specified)*

The List Data dialog box asks the user to fill in the variables desired. If you do not fill in any variables, all variables are listed. You can also choose to have a subset of the data listed by specifying the first and last case numbers to be listed.

# Basic Statistics

Means, standard deviations, medians, percentiles, and indicators of skewness and kurtosis are basic statistics that should be computed for most data sets as a means of checking on the integrity of the data.

## Means and Standard Deviations

StataQuest will output means and standard deviations for all or part of the data set by choosing *Means and SDs* from the *Summaries* menu. (SD stands for standard deviation.) This command produces several univariate statistics, including the mean or average, the standard deviation (defined for a sample with n-1 in the denominator; see any basic statistics text), the minimum value in the data set, and the maximum. These statistics are useful in doing the following:

- Assessing whether the data have been entered correctly. Excessively large or small values in the context of the particular data set often indicate errors.

- Assessing whether there are cases that are extreme on a given variable(s), though choosing *Medians/Percentiles* from the *Summaries* menu is better for this purpose.

- Seeing how many missing data points there are for each variable, in particular whether there are variables on which there are so few cases that we cannot use them in calculating any statistics.

- Obtaining the mean and standard deviation of a variable for use in other statistical calculations or comparing the distributions of different variables through the use of their standard deviations.

Figure 4.6 contains the means and standard deviations for the first six variables after the identification variable in the United Nations Human Development Report data set. These were obtained by opening the data set *humandev*, and selecting *Means and Sds* from the *Summaries* menu, and then specifying the variables to be computed in the dialog box. The DOS version of SQ works similarly.

In Figure 4.6, *land area* has only 128 observations, indicating missing data in two observations. Among the five other variables, there are no missing data points. Given that the mean population in 1960 is a little less than

```
┌─────────────────────────────────────────────────────────────────────┐
│ Stata Results                                                         │
├─────────────────────────────────────────────────────────────────────┤
│                                                                       │
│  . summarize pop1960 pop1988 urbpop60 urbpop88 landarea life          │
│                                                                       │
│  Variable |     Obs       Mean    Std. Dev.      Min       Max        │
│  ---------+-----------------------------------------------------      │
│   pop1960 |     130   23.66846    73.38908        .1       657        │
│   pop1988 |     130   39.23538    124.5581       1.1      1105        │
│  urbpop60 |     130   33.33077    24.05272         2       100        │
│  urbpop88 |     130   49.16154     24.7774         5       100        │
│  landarea |     128   100.4219    256.0489         0      2227        │
│      life |     130   62.71538    10.63287        42        78        │
│                                                                       │
└─────────────────────────────────────────────────────────────────────┘
```

**Figure 4.6**  *Means and Standard Deviations*

24 million, it is interesting that the largest value is 657 million, and the same phenomenon is true in 1988, indicating a possibly skewed variable. The percent urban population appears to be less skewed.

## Means and Standard Deviations by Group

It is possible to produce means, standard deviations, and the number of observations for groups within a data set. You could, for example, compare the average life expectancy for three groups of countries—the developed world, the newer developing countries, and the third world—in the U.N. Human Development data set. Or you could compare the average campaign expenditures for Republican and Democratic members of Congress. Here we will do the first, comparing life expectancies for different groups of countries.

## One-Way Means and Standard Deviation

Continuing with the U.N. Human Development Report data set provided with StataQuest, we want to compare the average life expectancy in different groups of countries. The life expectancy variable is called *life;* the variable dividing the nations into three groups is called *world*.

**1**  From the *File* menu, choose *Open* to read the *humandev.dta* data file from disk to memory if it is not there already.

**2**  From the *Summaries* menu, choose *Means and SDs by Group* and then *One-way of means*.

**3**  Within the "One-way Tabulation of Means" dialog box, select the "summary variable," in this case *life,* and the "group variable," in this case *world*.

```
  Stata Results

. tabulate world, summ(life)

Third world,| Summary of Life Expectancy (Years) 1987
 developing,|
       OECD|        Mean     Std. Dev.       Freq.
------------+------------------------------------------
   1-3rdWld |    59.103093    9.5976175          97
   2-Devel  |    70.533333    5.9265344          15
   3-OECD   |    75.666667    1.6803361          18
------------+------------------------------------------
      Total |    62.715385    10.632869          130
```

**Figure 4.7** *Mean Life Expectancy for Three Groups of Nations*

The result is shown in Figure 4.7.

We have three groups of nations, with the third world nations in category 1 having a mean life expectancy of only 55.2, compared with 67.7 and 75.0 for the newly industrializing nations in category 2 and the developed world (members of the Organization for Economic Cooperation and Development) in category 3. Note the difference in the standard deviations; group 3 seems much more homogeneous than groups 1 or 2.

## Two-Way Means

Two-way means is similar to one-way, except that instead of computing means within categories of one variable, we now do it within joint categories of two variables. (For a discussion of cross-tabulation, see Chapter 5.) Normally a cross-tabulation contains the number of observations and percentages of the total in each cell, but here we have the mean of a third variable in each cell. We illustrate with a data set of Congressmen, a 25% sample of the U.S. House of Representatives in 1994.

The data set *hrsamp94.dta* contains the results of three votes, as well as a variable for party affiliation. Let us compare the average amount spent on the representative's campaign in 1994 (*expwin*) by categories of political party (*party*) and whether or not the representative supported a bill to institute term limits on members of the House of Representatives (*termlim*).

**1** From the *File* menu, choose *Open* to read the *hrsamp94.dta* data file from disk to memory.

**2** From the *Summaries* menu, choose *Means and SDs by Group* and then *Two-way of means.*

**3** Within the Two-way Tabulation of Means dialog box, specify *party* as the row variable, *termlim* as the column variable, and *expwin* as the summary variable.

The result is shown in the top half of Figure 4.8.

We note that the tendency is for Democrats to spend more than Republicans regardless of their position on the bill and for Republicans who supported term limits to spend almost $100,000 more on their campaigns than Democrats who opposed the bill. We know from general knowledge that the bill was strongly supported by Republicans and opposed by Democrats. From the frequencies in each cell, we note that the vote was a strong party-line vote, but not totally overwhelming. Six Republicans opposed the bill, and 10 of the 46 Democrats in the sample supported it.

## Stata Results

```
. tabulate party termlim, summ(expwin)

Means, Standard Deviations and Frequencies of Winner Campaign
Spending 94

   Political | Vote on Term Limits 1995
       Party |        No           Yes          Total
-------------+--------------------------+----------
         Dem | 697.80833        695.99 | 697.41305
             | 505.99925     354.94392 | 473.64136
             |        36            10 |        46
-------------+--------------------------+----------
         Rep | 435.23334     513.68727 | 505.97049
             | 126.12209     251.22969 | 242.25038
             |         6            55 |        61
-------------+--------------------------+----------
       Total | 660.29762     541.73385 | 588.2729
             | 478.70111     274.52696 | 370.84005
             |        42            65 |       107
```

**Figure 4.8** *Means, Standard Deviations, and Numbers of Cases of Campaign Expenditures,1994, U.S. House of Representatives, by Party and Whether the Congressman Supported a Bill to Institute Term Limits on the U.S. House of Representatives*

## Median and Percentiles

The *Summaries* menu will not only provide us with means and standard deviations, it will also provide us, for a specific variable, with the median, several percentiles, the smallest four and the highest four values, and measures of skewness and kurtosis. These are useful as follows:

- **Median.** The median is used as a measure of central tendency that is unaffected by the extremes, as well as a statistic used to compare with the mean to determine whether a variable is skewed.

- **Percentiles.** To define a percentile, start by assuming we have whole numbers X from 1 to 99. Then the Xth percentile for any group of numbers is any value that has X% or less of the observations less than that value and (100-X)% of the observations greater than that value. So the 20th percentile has 20% of the observations less than that value and 80% greater than that value. Percentiles are useful for educators or psychologists in looking at standardized tests and for others who may want to form box plots or know the boundaries of computer-produced box plots. The median is the 50th percentile.

- **Four smallest and four largest values.** These are useful in seeing extreme points and in comparing those values with the 99th, 95th, or 90th percentile.

- **Number of observations.** This is the number of observations that are not missing data.

- **Sum of weights.** If the data are weighted, the sum of the weights on the non-missing data of a particular variable is useful in assessing whether there are problems in how the weights are operating and interacting with missing data.

- **Mean and standard deviation.** These can be compared with the median to see whether the data are skewed. The standard deviation is the mean's measure of dispersion.

- **Variance.** The variance is the standard deviation squared.

- **Skewness coefficient.** If the data are symmetric, this coefficient will be zero or close to zero. Negative values indicate that the data are skewed down, toward lower values. Positive values indicate upward skew, toward higher values. Judgments about a cause for concern vary by discipline. Ask your instructor for guidance.

- **Kurtosis coefficient.** This coefficient indicates the height or "peakedness" of the distribution in comparison with the normal distribution. The normal distribution has a kurtosis coefficient of 3. Higher values indicate more peakedness; lower values indicate that the distribution is flatter than the normal distribution.

The U.N. Human Development Report contains a variable for the population of each nation in 1988. We shall compare the 1988 population in its raw form (*pop1988*) with the variable transformed with the log transformation (*logpop88*). To do this:

1   From the *File* menu, choose *Open*.

2   Type the name of the file: humandev.

3   From the *Summaries* menu, choose *Median/Percentiles*.

4   **Windows:** Click on the variable named *pop1988*.

    **DOS:** Enter the variable named pop1988.

5   Repeat the process with the variable *logpop88*.

The results appear in Figure 4.9.

Because we know that the characteristic of variables like population, Gross National Product, income, and the like in data sets consisting of nations is that some of the variables can be quite positively skewed, we look first at the comparison of the mean and median. The median is the 50% figure. For *pop1988*, the median is 9.7 million people, whereas the mean is more than 39 million. When we look at the figures for skewness and kurtosis, we notice that the skewness figure is almost 7 and the kurtosis figure is more than 54, both far above the levels that we would get for this variable if it were distributed similarly to a "normal" distribution. The normal distribution has a skewness coefficient of 0 and kurtosis of 3.

We then look at the variable that has been modified by the log transformation. The variable has been transformed by the log transformation to the base 10. StataQuest's log function, however, is base e. To use the function and obtain a transformation to the base 10, you transform by the log function and divide by the log of 10, as follows: The formula for the new variable, *logpop88*, would be log(pop1988) / log(10). Its median and mean are quite close together at 0.99 and 1.06, and the skewness coefficient is now 0.68 instead of 6.89. The kurtosis coefficient is now closer to 3 also. So the logged variable is more similar to a normal distribution than the untransformed raw data.

## Confidence Intervals

StataQuest will produce a 95% confidence interval on any variable from your data set. Note that you calculate confidence intervals based on data in memory through the *Summaries* menu, or you calculate a confidence interval through the Statistical Calculator (the *Calculator* menu), entering the mean, standard deviation, and sample size directly from the keyboard (see Chapter 15). We will produce a confidence interval around the average age of the adult American

```
┌─────────────────────────────────────────────────────────────────────────┐
│ ████ Stata Results ████████████████████████████████████████████████████ │
├─────────────────────────────────────────────────────────────────────────┤

  . use C:\SQDATA\HUMANDEV.DTA, clear
  . summarize pop1988, detail

                     Population in Millions, 1988
  -----------------------------------------------------------------
              Percentiles      Smallest
    1%           1.1             1.1
    5%           1.5             1.1
   10%           2.2             1.2
   25%           4.2             1.2       Obs                    130
                                          Sum of Wgt.            130

   50%           9.7                       Mean             39.23538
                               Largest     Std. Dev.        124.5581
   75%          24              245
   90%          60              284        Variance         15514.71
   95%         122              820        Skewness         6.894673
   99%         820             1105        Kurtosis         54.12938

  . summarize logpop88, detail
                     Base-10 Logs of Pop in 1988
  -----------------------------------------------------------------
              Percentiles      Smallest
    1%          0.04            0.04
    5%          0.18            0.04
   10%          0.34            0.08
   25%          0.62            0.08       Obs                    130
                                          Sum of Wgt.            130

   50%          0.99                       Mean                 1.06
                               Largest     Std. Dev.            0.59
   75%          1.38            2.39
   90%          1.78            2.45       Variance             0.35
   95%          2.09            2.91       Skewness             0.68
   99%          2.91            3.04       Kurtosis             3.50
```

**Figure 4.9** *Median and Percentiles for* pop1988 *and* logpop88

public, as measured by the National Opinion Research Center's General Social Survey of 1994. The data set *gsssurv.dta* contains a 20% sample (598 observations) for six variables from the 1994 General Social Survey. The *age* variable asks the respondent's actual age in 1994 and is a measure of the age of the non-institutionalized adult (only persons 18 and older are surveyed) population. You could compare this result with the results of the much larger Current Population Survey for the same time period or with the last Census.

**1** From the *File* menu, choose *Open*.

**2** Enter the name of the file: gsssurv.

**3** From the *Summaries* menu, choose *Confidence intervals*.

**4** **Windows:** Click on the variable named *age*.

   **DOS:** Enter the variable named age.

The results appear in Figure 4.10.

---

**Stata Results**

```
. ci age, level(95)

Variable |    Obs      Mean    Std. Err.    [95% Conf. Interval]
---------+-------------------------------------------------------
     age |    596    45.29027   .6832032    43.94849    46.63205
```

**Figure 4.10** *Confidence Interval Around the Mean*

---

The average age is (rounding all numbers to one decimal place) 45.3, with a 95% confidence interval around the mean ranging from 43.9 to 46.6. We can be 95% confident that the true population mean lies between these two numbers, or, more formally, if we were to take a large number of samples of size 598 and calculate confidence intervals in each, 95% of these confidence intervals would contain the true population mean.

## ■ EXERCISES

**4.1** What is the source of the data in *transit.dta*? Do you have any questions about the variables and how they are coded after looking at the data and descriptive information?

**4.2** What is the source of the data in each of the following data sets? How is each variable coded? Do these numbers make sense to you? *elderly.dta, gsssurv.dta, husbands.dta, draft70.dta, humandev.dta, hrsamp92.dta.*

**4.3** Use *Means and SDs by Group* (within the *Summaries* menu) to compute the following:

   a. The average diameter of ball bearings by production line (*bearings.dta*).

b. The average amount of campaign expenditures by the winners and losers in the 1992 congressional races by party (*hrsamp92.dta*).

c. The average birth weight of babies in a low birthweight study by whether the mother smoked and whether the mother had a history of hypertension (*lwbtwt1.dta*).

d. The average number of days since the first hospitalization of terminal cancer patients in a study to determine the effects of Vitamin C, by the kind of cancer and whether the observation is a treatment case (received vitamin C) or control (*pauling.dta*).

In each case, find the requested average by each variable individually and then by the joint distribution of the two variables mentioned.

**4.4** The data set *wdr93.dta* is composed is five variables from the 1993 World Development Report, published by the World Bank. Form a new variable called *loggnp91* which is the log of *gnpcap91*. Do the same with *illit90*, calling the new variable *logill90*. Save the revised data set under the name *adc4p4*.

# Tables: Frequency Distributions and Cross-Tabulations

## What are Frequencies and Cross-Tabulations?

Data come in several varieties. One of the most useful classifications is between

- **Measurement variables.** Variables where individual data points can take on any value within a continuous interval on a number line. Examples: your age, anything that is measured in money, your height and weight, the Gross National Product, and so on.

- **Categorical variables.** Variables that are discrete, where individual data points are numbers that are distinct and countable. On a number line, categorical variables can only take on certain numbers, usually integers. Categorical variables can have two categories, such as success and failure, voted or didn't vote, male or female, to name only a few examples. Other categorical variables have more categories, such as the religion of the respondents in a national survey, or the political party they identify with (strong Democrat, weak Democrat, independent leaning Democrat, independent, independent leaning Republican, weak Republican, strong Republican, apolitical).

| | ■ | **Frequency distributions** |
| ■ ■ ■ ■ ■ ■ ■ | | |
| **IN THIS CHAPTER** | ■ | **Cross-tabulation** |
| | ■ | **Cross-tabulation by the categories of a third variable** |

The techniques of this chapter are directed toward categorical variables. What happens when you form a frequency distribution from a measurement variable will be illustrated in the next section.

## Frequency Distributions

### Categorical Variables

A frequency distribution is a table that tells you how data are grouped on the selected variable, that is, how many observations fall into each category. StataQuest will take a variable like *gender* or *ethnicity* and tell you how many persons fall into each of the several classifications. For *gender*, you can obtain the number and percentage of males and females. For *ethnicity*, you can obtain the number and percentage of African-Americans, Caucasians, Latinos, and so on.

We will use the 1994 General Social Survey, a random sample of the American public contained in data set *gsssurv.dta* to do frequency distributions of several variables.

1 From the *File* menu, choose *Open*.

2 Type the name of the file: `gsssurv`.

3 From the *Summaries* menu, choose *Tables* and then *One-way (frequency)*.

4 **Windows:** Click on the variable named *educ*.

   **DOS:** Enter the variable named `educ`.

The results are shown in the top half of Figure 5.1.

In the top half of Figure 5.1, education has been recoded into two categories, 0 for those with a high school diploma or less, and 1 for those with a more than a high school diploma. We note that there are 305 observations or 51.35% of the total in the first category, and 289 or 48.65% of the sample in the second category.

```
Stata Results

. tabulate educ
Respondent's|
  education|       Freq.       Percent        Cum.
-----------+---------------------------------------------------------------
  0<HighSc |         305       51.35        51.35
     1>HS  |         289       48.65       100.00
-----------+---------------------------------------------------------------
    Total  |         594      100.00

. tabulate educ, plot
Respondent's|
  education|Freq.
---------+------+-------------------------------------------------------
0<HighSc |  305 |********************************************************
    1>HS |  289 |********************************************************
---------+------+-------------------------------------------------------
  Total  | 594
```

**Figure 5.1** *Frequency Distribution of Education*

The bottom half of the figure shows the result of checking the box labeled "Include character plot" (Windows only). A character histogram is listed on screen and will become part of your log file if one is open. But with this presentation of the data, there are no percentages computed.

Notice that you must specify each variable **individually** in running frequency distributions.

## Measurement Variables

A frequency distribution on a measurement variable often results in only one or two observations falling into each category, like the *fare* variable from the *transit.dta* data set, shown in Figure 5.2.

Similar results can be obtained from the age variable in the *gsssurv.dta* data set used in Figure 5.1. **Measurement variables need to be categorized using the recode command** illustrated in Chapter 3 so that they change from being **continuous variables**, where any value can be valid, to **discrete or categorical variables**, where only a few values are valid. In the General Social Survey example, we had already recoded the education variable from the number of years of education to 0 meaning less than 12 years of education and 1 meaning 12 years or more. The *budget* variable from the *transit.dta* data set could be recoded into

```
Stata Results

. tabulate fare
Minimum Fare|
     Per Ride|     Freq.        Percent        Cum.
-----------+-----------------------------------------------
      0.75 |       1            10.00         10.00
      0.85 |       2            20.00         30.00
      1.00 |       2            20.00         50.00
      1.10 |       2            20.00         70.00
      1.15 |       1            10.00         80.00
      1.25 |       1            10.00         90.00
      1.50 |       1            10.00        100.00
-----------+-----------------------------------------------
     Total |      10           100.00
```

**Figure 5.2** *Frequency Distribution of a Measurement Variable*

those systems that spend less than 500 million dollars per year and those that spend more, or, alternately, three categories: less than $500 million, $501 to $1,000 million, and over $1,000 million. Any **values** could be used for this categorization, but it is customary to use integers such as 1, 2, and 3.

## Cross-Tabulation

Tables can tell you the number of observations for each category of a second variable that fall into each category of a first variable. For example, with the General Social Survey of 1994, we have the following categorical variables: education (*educ*), ethnicity (*ethnicit*), income (*income*), gender (*gender*), and whether the respondent voted in 1992 (*vote92*). With *Tables' Two-way (crosstabulation)* procedure, we can tell how many persons of each gender voted or not; how many poor people voted compared with how many relatively not-poor, and so on.

In doing a cross-tabulation, you must specify which variable is the row variable and which is the column variable. The normal convention is the following:

**Row variable:** Variable of interest; could be a "dependent" variable.
**Column variable:** Independent variable, one that relates to the variable of interest in some way.
**Percentages:** You use percentages of the column variable and compare them across rows (see Bowen and Weisberg, 1980, 60–66).

We shall produce a table where the variable of interest is whether the respondent voted in 1992; the related variable is education. To produce the table, we take the following steps:

**1** From the *File* menu, choose *Open*.

**2** Enter the name of the file: gsssurv.

**3** From the *Summaries* menu, choose *Tables* and then *Two-way (cross-tabulation)*.

**4** **Windows:** Click on *vote92* for the row variable; *educ* as the column variable; click on *column percentages* as per the directions above. Do not check the box labeled "Report all statistics."

   **DOS:** Enter the row variable name: vote92; enter the column variable name: educ; answer yes to the general question on percentages and yes to column percentages. Answer no to all other questions.

The results are shown in Figure 5.3.

```
Stata Results

. tabulate vote92 educ, chi2 column

Did R vote| Respondent's education
  in 1992|
 election?|   0<HighSc       1>HS |     Total
-----------+----------------------+------------------------------
 0Abstain |       111         56 |       167
          |     38.01      20.07 |     29.25
-----------+----------------------+------------------------------
 1-Voted  |       181        223 |       404
          |     61.99      79.93 |     70.75
-----------+----------------------+------------------------------
    Total |       292        279 |       571
          |    100.00     100.00 |    100.00

        Pearson chi2(1) = 22.1956 Pr = 0.000
```

**Figure 5.3** *Cross-Tabulation of Whether the Respondent Voted in 1992 (*vote92*) by Education (*educ*)*

---

▌**TIP 6** ▌ ▌ ▌ ▌ ▌ ▌ ▌ ▌ ▌ ▌ ▌ ▌ ▌ ▌ ▌

## Row versus Column Percentages in Cross-Tabulations

If you receive the wrong percentages, simply rerun the cross-tabulation. Or always obtain both row and column percentages.

---

### Percentage Difference Interpretation

The results show that 38.01% of the lesser educated group failed to vote, compared with 20.07% of the better educated group. Almost 62% of the less educated group voted, compared with almost 80% of the better educated group.

### Chi-Squared Interpretation

A chi-squared test of the null hypothesis that the two variables are independent and thus unrelated results in a chi-squared of almost 22.2, with the probability of obtaining a result this extreme or more by chance is less than 0.0005. (The probability listed is 0.000. If the probability were greater than or equal to 0.0005, that probability would round to 0.001. Thus the statement that the probability is less than 0.0005.) We leave a more formal interpretation of the null hypothesis of a chi-squared statistic to your statistics textbook.

### Other Statistics

StataQuest is capable of computing several other statistics traditionally used in analyzing cross-tabulations. These are computed by (Windows) checking the box labeled "Report all statistics" in the Two-way (cross) tabulation dialog box, or (DOS) answering the question about reporting all statistics with a "yes." The particular statistics are the following:

- **The likelihood-ratio chi-squared**—Similar to the chi-squared statistic in having degrees of freedom equal to the number of rows minus one times the number of columns minus one. It uses the natural log of the number expected in each cell instead of the sum used by chi-squared. Like chi-squared, it is appropriate for all tables.

- **Cramer's V**—A chi-squared based statistic that varies between -1 and +1 for 2 by 2 tables and between 0 and 1 otherwise. Like chi-square, it is appropriate for all tables.

- **Gamma**—This statistic is appropriate for tables where both variables are ordered, that is, they have ranks. Gamma varies between -1 and +1. A common interpretation is for a "low" correlation to be between 0 and .3; a "moderate" correlation from .3 to .6; a "high" correlation, greater than .6. The same holds true for negative correlations. See your statistics text for more information.

■ **Kendall's tau-b**—This is appropriate for tables where both variables are ordered, that is, they have ranks. It varies between -1 and +1. The same interpretation of the range of a correlation applies as for gamma, above.

In Figure 5.4, we have a highly significant chi-squared, meaning that we can reject the hypothesis that there is no relationship between the two variables. The table is seemingly not random. Gamma is at the high end of the low-to-moderate correlation range; Kendall's tau-b is at the low end of the same range (it is often the case that gamma correlations will be considerably greater than Kendall's tau; Kendall's tau tends to yield similar results to Pearson's r; Kendall's tau also tends to be less sensitive than gamma to the number of response categories).

---

## Stata Results

```
. tabulate vote92 educ, chi2 column all

Did R vote| Respondent's education
   in 1992|
 election?|  0<HighSc        1>HS |     Total
----------+---------------------------+-----------------------------------
  0Abstain |      111          56 |       167
           |    38.01       20.07 |     29.25
----------+---------------------------+-----------------------------------
  1-Voted  |      181         223 |       404
           |    61.99       79.93 |     70.75
----------+---------------------------+-----------------------------------
     Total|      292         279 |       571
           |   100.00      100.00 |    100.00

            Pearson chi2(1) =   22.1956   Pr = 0.000
   likelihood-ratio chi2(1) =   22.5346   Pr = 0.000
                 Cramer's V =    0.1972
                      gamma =    0.4190   ASE = 0.079
            Kendall's tau-b =    0.1972   ASE = 0.040
```

**Figure 5.4** *Additional Statistics Reported for the Cross-Tabulation in Figure 5.3*

# Cross-Tabulations by the Categories of a Third Variable

In many situations data analysts want to examine a two-variable table for the effects of a third variable. What will happen, for example, to the relationship between education and the tendency to vote in 1992 if we consider the effects of income? Perhaps the real reason that more highly educated people have a higher tendency to vote is because they are richer. When you analyze a cross-tabulation by the categories of a third variable, you hold that third variable constant while you see whether the relationship between the first two variables persists. In this case, we will look within each category of the income variable to see whether the relationship between education and the tendency to vote is present.

To obtain a three-way cross-tabulation,

**1** From the *File* menu, choose *Open*.

**2** Type the name of the file: `gsssurv`.

**3** From the *Summaries* menu, choose *Tables* and then *Three-way (by group)*.

**4** **Windows:** Click on *vote92* for the row variable; *educ* as the column variable; and *income* as the "By variable." Click on "Report column percentages" and "Report all statistics" again.

**DOS:** Enter the row variable name: `vote92`; enter the column variable name: `educ`; then answer `yes` to the questions on column percentages and reporting all statistics; answer `no` to all other questions; the variable identifying the groups is `income`.

The results are shown in Figure 5.5.

Some notes on Figure 5.5:

■ **Part of the table is missing.** A third cross-tabulation printed out, for those people for whom the income variable was missing data (income = . ). In most surveys in the U.S., a substantial portion of respondents commonly refuse to report their incomes; we shall ignore these persons because they did not provide their incomes and because doing so enables the other two cross-tabs to fit on one page. We have eliminated the line with "Cramer's V" statistic for the same reason.

■ **Percentage Differences:** Previously we noted a substantial percentage difference between those with less and those with more education; those with more education had a higher probability of voting. Does this relationship hold up for these two cross-tabulations? Looking at the percentage of those who voted, we see for the first cross-tab where income is low

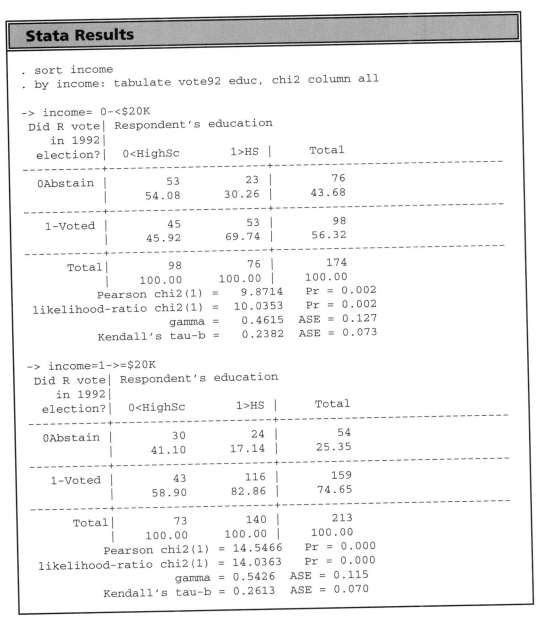

```
Stata Results

. sort income
. by income: tabulate vote92 educ, chi2 column all

-> income= 0-<$20K
 Did R vote| Respondent's education
    in 1992|
  election?|  0<HighSc      1>HS |      Total
-----------+----------------------+---------------------------------
   0Abstain |        53         23 |         76
           |     54.08      30.26 |      43.68
-----------+----------------------+---------------------------------
    1-Voted |        45         53 |         98
           |     45.92      69.74 |      56.32
-----------+----------------------+---------------------------------
      Total|        98         76 |        174
           |    100.00     100.00 |     100.00
           Pearson chi2(1)  =    9.8714   Pr = 0.002
   likelihood-ratio chi2(1) =   10.0353   Pr = 0.002
                    gamma =     0.4615   ASE = 0.127
          Kendall's tau-b =     0.2382   ASE = 0.073

-> income=1->=$20K
 Did R vote| Respondent's education
    in 1992|
  election?|  0<HighSc      1>HS |      Total
-----------+----------------------+---------------------------------
   0Abstain |        30         24 |         54
           |     41.10      17.14 |      25.35
-----------+----------------------+---------------------------------
    1-Voted |        43        116 |        159
           |     58.90      82.86 |      74.65
-----------+----------------------+---------------------------------
      Total|        73        140 |        213
           |    100.00     100.00 |     100.00
           Pearson chi2(1) =   14.5466   Pr = 0.000
   likelihood-ratio chi2(1) =   14.0363   Pr = 0.000
                    gamma =    0.5426   ASE = 0.115
          Kendall's tau-b =    0.2613   ASE = 0.070
```

**Figure 5.5** *Three-Way Cross-Tabulation of Education by Whether the Respondent Voted in 1992, Controlling for the Respondent's Income*

that almost 70% of those with more education voted versus almost 46% of those with less education. For the second cross-tab, where income is higher, those two percentages are 83% versus 59%. So in each category, a higher percentage votes when income is higher, but the difference between those with more education and those with less education persists. It doesn't disappear.

■ **Statistics:** Another way to see if income has made a difference is to look at the statistics we analyzed earlier for the two-variable table. Chi-squared is still significant in each cross-tab; both gamma and Kendall's tau-b are still in the same general range.

■ **Conclusion:** We conclude that the low-to-moderate relationship between education and the tendency to vote persists even when we control for income.

For more information, the reader is referred to the social science literature on spurious relationships and controlling for third variables. Spurious relationships are those which are due to the effects of an omitted third variable. Many analysts prefer to control for these relationships through multiple regression and other advanced techniques. For more information on spurious relationships and cross-tabs, see Bowen and Weisberg.

## ■ EXERCISES

Hint: each of the cross-tabulation problems is easier if you run *Describe variables, Dataset information*, and perhaps *List variables* from the *Summaries* menu, as well as a frequency distribution on each variable, prior to running the cross-tabulation.

5.1 *hrsamp94.dta* is a 108 observation random sample of the 1995 U.S. House of Representatives, including data on each Representative's party affiliation (Republican or Democratic), and three votes: *brady* (the vote on the Brady gun control bill), *nafta* (the vote on the North American Free Trade Alliance), and *termlim* (the vote on the bill to institute term limits on the House of Representatives). Run three cross-tabulations, using *brady*, *nafta*, and *termlim* as the row variables in turn and *party* as the column variable. Use column percentages. Which of the three variables has the strongest split between the Republicans and Democrats? Use percentage differences as the basis for deciding which variable has the strongest split, and interpret each of the three tables in 1-3 sentences.

**5.2**    Data sets *draft70.dta* and *draft71.dta* contain the results of the first two random selections of draft numbers. Compare 1970 and 1971 with a cross-tab of *month* by *dfthilo1* (1970) and *month* by *dfthilo2* (1971). Use row percentages for each. Are the percentages approximately what you would expect from a random drawing? Interpret each table in 2–3 sentences. Be sure you understand from an examination of the data set exactly what each observation and variable represent.

**5.3**    Compare cross-tabulations of gender (the variable *sex*) by the different confidence in institutions variables included in *gsssurv2.dta*, a cross-section of the American public in 1994. In which institution(s) is there a marked difference across the genders? In which is there not such a difference? For this exercise *sex* should be the row variable, *conbus*, *coneduc*, *conlegis*, *consci*, and *contv* should be the column variables in turn. Use column percentages. Interpret each cross-tabulation in 1–2 sentences.

**5.4**    This exercise uses 3-way cross-tabulations. Use the data file *capital.dta* to examine the disposition of capital punishment cases in the State of Florida. First, do two 2–way cross-tabulations: *rk* by *rv* and *rk* by *death*. Interpret each in 2–3 sentences. Then run *rk* by *death* for the categories of *rv*. Again, interpret the result. *capital.dta* is a study of a decade's worth of capital punishment cases in the State of Florida.

**5.5**    Use the file *gsssurv.dta*, a random sample of the 1994 American public, to ascertain which variable is the strongest predictor of whether the respondent voted in 1992. Use *vote92* for the row variable and *educ*, *ethnicit*, *income*, and *gender* in turn as column variables to produce four 2x2 tables. Use column percentages. Interpret each in 1–3 sentences. Using percentage differences or any appropriate statistics, which variable is the best single predictor?

# 6

# Histograms and
# Normal Quantile Plots

## Graphs for Studying the Shapes of Distributions

**Histograms** show the frequency distribution of a variable. They display bars with heights proportional to the fraction or number of observations found at each value. If the variable is continuous or has a wide range of discrete values, it will be necessary to group values on the scale into a smaller number of fixed intervals or "bins" for counting purposes. Values with relatively few observations will have shorter bars; those with more observations will have taller bars. The appearance of how these bars of varying heights line up across the scale is what we mean by the "shape" of a distribution. A distribution can have many shapes. For example, it can be bell-shaped or rectangular, positively skewed or negatively skewed, bimodal or multimodal, fat-tailed or thin-tailed, smooth-looking or full of gaps, and occasionally just plain weird.

Many statistical procedures assume that a variable has a bell-shaped normal distribution. It is important to check the validity of this assumption. One StataQuest tool for doing this is the optional **normal curve overlay** on histograms. The closer this overlay matches the actual shape of the histogram,

▓ ▓ ▓ ▓ ▓ ▓ ▓

## IN THIS CHAPTER

- ■ **Histograms in DOS versus Windows**
- ■ **Continuous variable histograms**
- ■ **Discrete variable histograms**
- ■ **Continuous variable histograms by group**
- ■ **Discrete variable histograms by group**
- ■ **Normal quantile plots**

the more confident you can be that the variable is normally distributed. The **normal quantile plot** is another graphical tool for assessing normality. This graph plots the actual percentile value of each observation against its expected percentile assuming a normal distribution. If the data points coincide with a line of equality, the variable may be normally distributed.

## Histograms in DOS versus Windows

The histograms displayed in SQ for DOS lack some of the options available in the Windows version. For example, SQ for Windows plots histograms separately for continuous and discrete variables; the DOS version doesn't distinguish between types of variables. In addition, although the DOS version provides options for the number of bins and a normal curve overlay (see the following), it lacks the range, bin width, draw, and auto-redraw options offered in the Windows version

To make a histogram in the DOS version, respond to the prompts by entering the name or number of the variable to be graphed, the number of bins desired, and `yes`, if you want a normal curve overlay. The text that follows applies to the Windows version, and notes are added where relevant for those using DOS.

## Continuous Variable Histograms

**Continuous variables** are quantitative variables that can be measured in fractional values and expressed in decimals to any level of precision desired. An example is temperature, which might be measured as 85 or 85.24 or 85.23944369 degrees Fahrenheit. **Discrete variables** are quantitative variables that can't be measured in fractional values and must be expressed as integers.

You might place 1 or 4 or 12 eggs in a basket, for example, but you can't place 5.281 eggs there. This section shows you how to make histograms of continuous variables. The next section discusses discrete variable histograms.

## Choices and Options

To make a continuous variable histogram, you need to group values across a **range** of the data into a set of fixed intervals of equal **width** called **bins**. Although you will almost always want to cover the full range of your data, there are no hard-and-fast rules for choosing the number of bins or interval widths. For example, to make a histogram of temperatures that range from 3.23 to 98.56 degrees Fahrenheit, you might group values into ten bins of width 10 degrees. The first bin would contain temperatures of 0 to 9.99 degrees, the second temperatures of 10.00 to 19.99 degrees, and so on, up to the tenth bin containing temperatures of 90.00 to 99.99 degrees. You would then count the number of temperatures that fall into each bin and make the histogram by drawing bar heights proportional to the counts. On the other hand, you might also choose to group values into twenty bins of width 5 degrees, five bins of width 20 degrees, and so on. Figure 6.1 shows four different histograms with varying numbers of bins and interval widths to illustrate the visual effects of such choices.

As shown in Figure 6.1, decisions about the number of bins and interval widths can dramatically affect the overall appearance of the histogram and its value as a visual aid. The two histograms at the top of Figure 6.1 clearly have too few bins to convey much detail about the shape and spread of this distribution. The two histograms at the bottom are much more informative.

Figure 6.2 shows StataQuest's dialog box for the continuous variable histogram, containing several option boxes and buttons that allow you to make these kinds of choices.

### Range From/To

SQ automatically sets the minimum and maximum on the scale to cover the full range of your data. Under Options, you can set your own minimum and maximum by typing the desired minimum in the "from" box and the desired maximum in the "to" box. For example, you might type 5 as the minimum in the "from" box and 50 in the "to" box. SQ will draw the histogram to include all values in your data from 5 to 50, but will not include any values below 5 or above 50. Restricting the range of a histogram is one way to "zoom" on a segment of the scale to study the distribution in finer detail.

### Number of Bins (Without Auto-Redraw)

Unless you indicate otherwise, SQ will assume that you want 8 bins for your histograms. The number 8 (or the last number entered) will be shown in the option box. There are two ways to change it to another number: (1) Click in

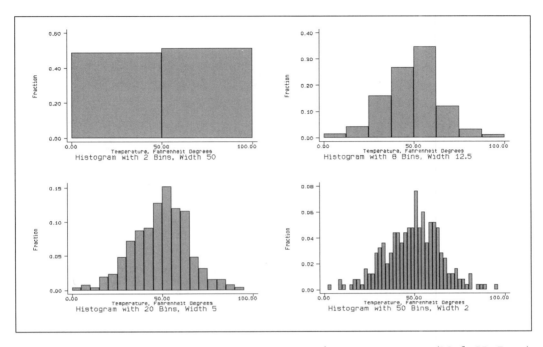

**Figure 6.1** *Four StataQuest Histograms of 250 Temperatures (Made-Up Data) Illustrating the Effects of Varying Numbers of Bins and Interval Widths*

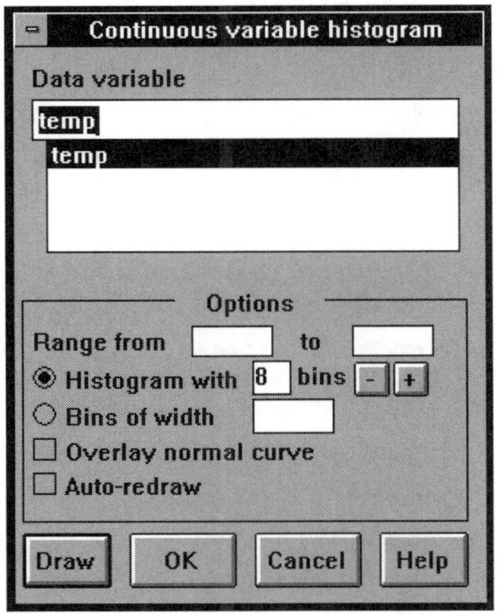

**Figure 6.2** *Histogram Dialog Box*

the box, select the text, and then type the new number to overwrite the old one. (2) Click once on the "+" button to add one more bin, click twice to add two more, and so on, up to **SQ's maximum of 50 bins**. Or click once on the "−" button to subtract one bin, click twice to subtract two, down to **SQ's minimum of 2 bins**. Then click on the Draw or OK button and a histogram with the desired number of bins will be displayed.

### Number of Bins (With Auto-Redraw)

Click on the *Auto-redraw* option and then click repeatedly on the "+" or "−" buttons to raise or lower the number of bins. The histogram displayed in the Graph Window will be instantly redrawn to reflect each change as you make it. In other words, with the *Auto-redraw* option, you don't have to click the Draw or OK button to see the result of each change made. This animated graphical display saves a lot of time in experimenting with different bin numbers to find one that seems optimal for your data.

One minor drawback of using *Auto-redraw*, however, is that the "+" and "−" toggle buttons are covered over by the Graph Window when it is brought to the front. You can click on the Dialog button to put the dialog box in front once again. A better solution, however, is to **push the Graph Window out of the way**. With the mouse, move your cursor to the left border of the Graph window. When the cursor turns into a double-arrow, click and drag the border toward the right just far enough to reveal the "+" and "−" buttons on the dialog box. Then release the mouse button. The Graph window will be a bit smaller than before, but now you can click the number of bins up or down and see instant results without interruption.

### Bin Width

If you want to specify the exact width of each bin in the histogram, click on the radio button to the left of *Bins of width*, which will deactivate use of the number of bins option. Then click on the box to the right and type the desired bin width. You may want to set the range before doing this. For example, you might set the range from 0 to 100 and then select 10 as your bin width. SQ will display a histogram with ten bins of width 10 on a scale ranging from 0 to 100. **If you ask for a bin width that will produce fewer than 2 bins or more than 50, SQ will give you an error message.** If that happens, try a different width.

### Overlay Normal Curve

Select the *Overlay normal curve* option if you want to see how closely the shape of your histogram matches the smooth outline of what it "should" look like if it is normally distributed. When you click on this option, SQ calculates the mean and standard deviation of the variable and uses these with a formula to overlay a normal curve on the histogram. The area under the normal curve is exactly equal to the area taken up by the histogram. If the contours of the histogram and overlaid curve largely coincide, the variable is normally distributed. If large chunks of the histogram stick outside the overlaid curve, however, the variable is not normally distributed. The visual comparison will help you see at a glance whether a distribution is positively or negatively skewed, has fat or thin tails, and so on.

## Three Examples of How to Make Continuous Variable Histograms

These examples use the *humandev.dta* file, which contains various data on 130 world nations. The three variables of interest are GNP per capita in 1987 (*gnpcap87*), percent adult literacy in 1985 (*literacy*), and percent living in urban areas in 1988 (*urbpop88*).

### Example #1: Histogram of 1987 GNP per Capita (10 Bins, Normal Curve Overlay)

**1**   From the *File* menu, choose *Open*.

**2**   Enter the name of the file: humandev.

**3**   From the *Graphs* menu, choose *One variable*, then *Histogram*, and then *Continuous variable*.

**4**   **Windows:** Click on the variable named *gnpcap87*. Click on the "+" button to select 10 bins. Click on the *Overlay normal curve* option.

   **DOS:** Enter the variable named gnpcap87. Answer prompts to select 10 bins and normal curve overlay.

The results appear in Figure 6.3.

This histogram shows that GNP per capita is severely positively skewed, with most observations piled up at the extreme low end of the scale and the rest straggling out toward the high end. The high GNP nations are probably outliers, as a box plot will confirm (see Chapter 8). The contrast with the normal curve overlay adds further visual evidence that this variable is not normally distributed.

### Example #2: Histogram of Adult Literacy 1985 (12 Bins, Normal Curve Overlay)

**1**   From the *File* menu, choose *Open*.

**2**   Enter the name of the file: humandev.

**3**   From the *Graphs* menu, choose *One variable*, then *Histogram*, and then *Continuous variable*.

**4**   **Windows:** Click on the variable named *literacy*. Click on the "+" button to select 12 bins.

   **DOS:** Enter the variable named literacy. Answer prompts to select 12 bins.

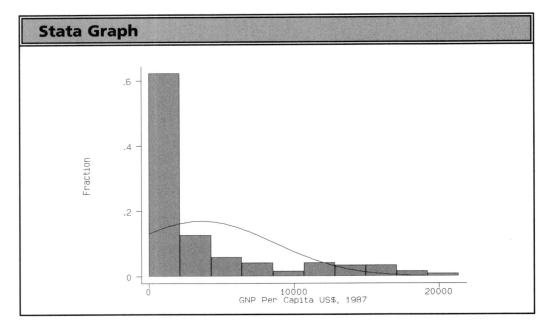

**Figure 6.3** *Histogram of 1987 GNP per Capita in 119 Nations*

The results appear in Figure 6.4.

This histogram shows that percent adult literacy is negatively skewed, with most observations piling up at the high end of the scale and the rest straggling out toward the low end. Note the empty bin at the extreme low end of the scale. The contrast with the normal curve overlay reinforces the impression that this variable is not normally distributed.

### Example #3: Histogram of Percent Urban 1988 (Range 0 to 100, Bin Width 20)

**1** From the *File* menu, choose *Open*.

**2** Enter the name of the file: humandev.

**3** From the *Graphs* menu, choose *One variable*, then *Histogram*, and then *Continuous variable*.

**4** **Windows:** Click on the variable named *urbpop88*. Enter 0 and 100 as the range. Enter 20 as the bin width.

**DOS:** The range and bin width options are not available. To produce the histogram shown in Figure 6.5, enter the variable named urbpop88 and answer prompts to select 5 bins.

The results appear in Figure 6.5.

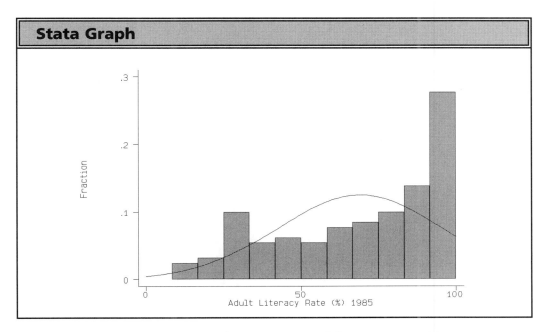

**Figure 6.4** *Histogram of 1985 Percent Adult Literacy in 130 Nations*

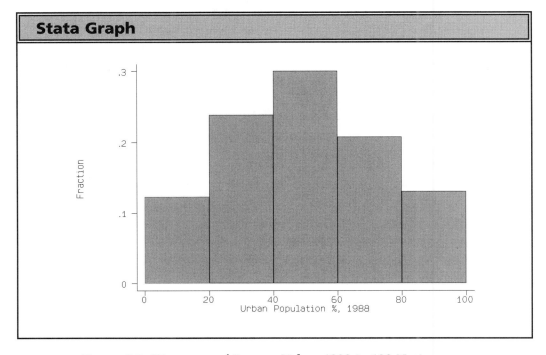

**Figure 6.5** *Histogram of Percent Urban 1988 in 130 Nations*

This histogram indicates that percent urban is approximately normally distributed. The overall center of the distribution is near the 50% mark. With 130 nations to plot, smaller bin widths would have helped to show greater detail. Note that specification of bin width 20 produced 5 bins to cover the selected range.

## Discrete Variable Histograms

SQ's discrete variable histogram will draw bars proportional to the number of observations for as many as 50 integer values in your data. The data must be integers and the range of values must not exceed 50. For example, integer data ranging from 1 to 50 can be plotted, as can data ranging from 120 to 169. The following kinds of data can't be plotted using this type of histogram: 2.34, 5.67, 12.4 (no decimals allowed, integers only) and 1, 2, 3, .., 52 (exceeds range limit of 50).

Continuous variables that are rounded to integers and that satisfy these conditions can be plotted as discrete variable histograms. Conversely, discrete variable data with a range of values exceeding 50 must be plotted using the continuous variable histogram.

There are no dialog box options available for this type of histogram. Click on the variable and then click on Draw or OK.

**Example:** The *cointoss.dta* file contains the results of 500 computer-simulated coin toss experiments. Each experiment involved tossing a fair coin 10 times and then recording the number of heads. Theoretically, the integer data can range from a minimum of 0 (all tails) to a maximum of 10 (all heads). Statisticians using the binomial theorem claim that the distribution will become approximately normal with increasing N. Let's check this claim visually with a discrete variable histogram.

**1** From the *File* menu, choose *Open*.

**2** Enter the name of the file: cointoss.

**3** From the *Graphs* menu, choose *One variable*, *Histogram*, and then *Discrete variable*.

**4** **Windows:** Click on the variable named *heads*.

 **DOS:** Enter the variable named heads, and then answer prompts with 11 bins and no normal curve overlay.

The results appear in Figure 6.6.

The histogram in Figure 6.6 indicates that the distribution of the discrete random variable *heads* is approximately normal for the 500 computer-simulated experiments studied here. (To learn how these computer simulations were done in SQ, open the *cointoss.dta* file, choose *Summaries*, and then *Dataset info.*)

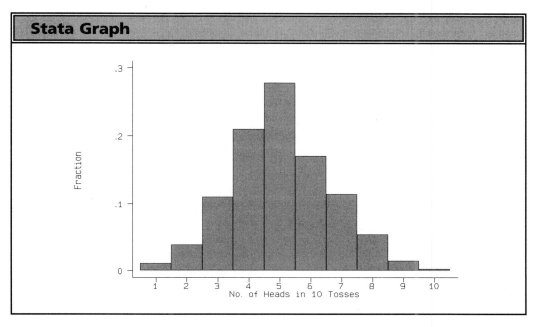

**Figure 6.6** *Discrete Variable Histogram of Number of Heads in 10 Flips of a Fair Coin. (N = 500 Computer Simulated Experiments Generated by a Binomial Distribution with p = .50)*

## Continuous Variable Histograms by Group

This menu command will display a continuous variable histogram for each value of a categorical group variable. All of the dialog box options discussed earlier are available, except for the options to set the range and bin widths. You can also request a histogram of the total in addition to those displayed for each group.

**Example:** In 1970, at the peak of the Vietnam War, claims were made that the Selective Service draft lottery that year produced nonrandom assignments of draft numbers (1, 2,...,366) to birthdays (1, 2, ..., 366), penalizing 18-year olds born in later months of the calendar year. The birthday numbers started at 1 for those born on January 1 and ended at 366 for those born on December 31, with a number for those born on February 29 in leap years included. Those who received lower draft numbers were called first to military service. The specific claim was that those with higher birthday numbers were significantly more likely to have received low draft numbers than those with lower birthday numbers. (See Williams, 1987; Feinberg, 1973.)

Using data for the results of the 1970 draft lottery (in the file *draft70.dta*), we will test this claim visually by displaying and comparing two continuous variable histograms of assigned draft numbers. The first will show the distribution of draft numbers assigned to 18-year olds with birthday numbers between 1 and 183 ("low" birthday group). The second will show the distribution of draft numbers assigned to those with birthday numbers between 184 and 366 ("high" birthday group).

**1**　From the *File* menu, choose *Open*.

**2**　Enter the name of the file: draft70.

**3**　From the *Graphs* menu, choose *One variable by group*, then *Histograms by group*, and then *Continuous variable*.

**4**　**Windows:** Click on the data variable named *dftno70* and the group variable named *day_hilo*. Click on the "+" button to set the number of bins to 15 (or type 15 in the box).

　　**DOS:** Enter the data variable named dftno70 and the group variable named day_hilo. Enter 15 as the desired number of bins.

The results appear in Figure 6.7.

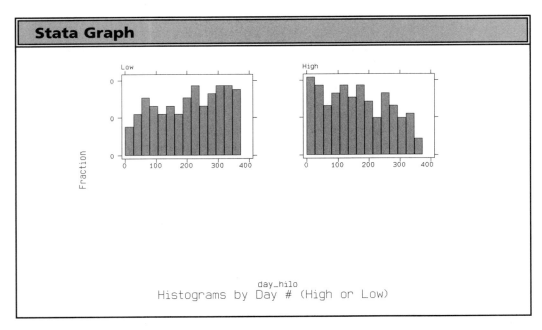

**Figure 6.7** *Histogram of Draft Numbers Assigned to Low and High Birthday Groups*

Comparing the two histograms shown in Figure 6.7, we can see clear visual evidence that the claim made has some merit. The distribution of draft numbers shown for 18-year olds with "low" birthday numbers is distinctly negatively skewed; that shown for 18-year olds with "high" birthday numbers is distinctly positively skewed. Allowing for some variation, we would expect both histograms to have a roughly rectangular shape if the draft lottery were truly random. Prompted by this kind of evidence, investigation revealed that the mechanical procedure used to make random draws was flawed. Changes were made in the design and mechanics of the lottery in 1971. (See Exercise 6.3.)

## Discrete Variable Histograms by Group

This menu command will display a discrete variable histogram for each value of a categorical group variable. You can also request a histogram of the total in addition to those displayed for each group. There are no other Dialog box options available.

**Example:** In the 1992 Presidential election, were there regional differences in the pattern of voter support for the incumbent candidate George Bush? In the *hrsamp94.dta* file, the variable named *bush92* gives the percentage vote for Bush in a sample of 108 Congressional districts. The categorical variable named *region* groups districts into four major regions: Northeast, Midwest, South, and West. The voting percentages are rounded to integers and range from 13 to 57. Thus, we can ask for discrete variable histograms of the percentage vote for Bush by the group variable *region*.

**1**  From the *File* menu, choose *Open*.

**2**  Enter the name of the file: `hrsamp94`.

**3**  From the *Graphs* menu, choose *One variable by group*, then *Histograms by group*, and then *Discrete variable*.

**4**  **Windows:** Click on the data variable named *bush92* and the group variable named *region*. Click on the *Include histograms of total* option.

**DOS:** Enter the data variable named `bush92` and the group variable named `region`. Enter 50 as the number of bins. (The histograms of total menu option is not available in the DOS version.)

The results appear in Figure 6.8.

A comparison of the histograms by region shown in Figure 6.8 suggests that Bush's overall pattern of voter support was strongest in the South and weakest in the Northeast. The four histograms display considerable district-to-district variation within each region. The histogram shown for the total of 108 districts is a composite of the four regional distributions. It indicates that for the nation as a whole Bush's percentage voter support was approximately normally distributed.

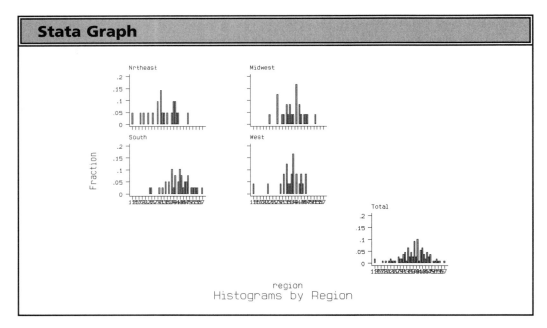

**Figure 6.8** *Discrete Variable Histograms of Percentage Vote for Presidential Candidate George Bush in 1992 in a Sample of 108 Congressional Districts, by Region*

## Normal Quantile Plots

The normal quantile plot is a useful tool for visually checking how closely the shape of a variable's distribution matches that of the normal ("bell-shaped") curve assumed in many statistical methods. Similar in function to the normal curve overlay option in the histogram command, the normal quantile plot allows the eye to see the degree of fit or misfit between ideal form and empirical reality. This graph plots the actual percentile value of each observation against its expected percentile, assuming a normal distribution. If the data points coincide with a line of equality, the variable is normally distributed.

**Example #1:** Let's make a normal quantile plot of the variable *urbpop88*, which gives the percentage of a nation's population living in urban areas in 1988. Is this variable normally distributed?

**1**  From the *File* menu, choose *Open.*
**2**  Enter the name of the file: humandev.
**3**  From the *Graphs* menu, choose *One variable,* and then *Normal quantile plot.*
**4**  **Windows:** Click on the data variable named *urbpop88.*
     **DOS:** Enter the data variable named urbpop88.

The results appear in Figure 6.9.

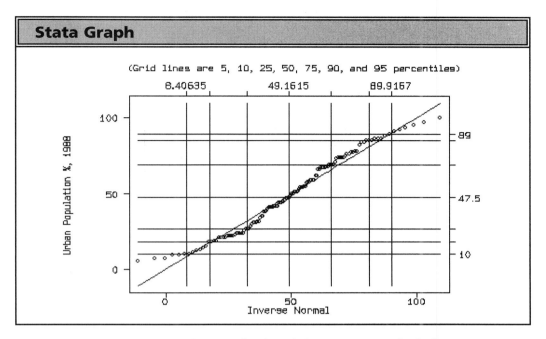

**Figure 6.9** *Normal Quantile Plot of the Percentage of Inhabitants Living in Urban Areas in 1988 in 130 Nations.*

The normal quantile plot shown in Figure 6.9 indicates that the variable *urbpop88* is approximately normally distributed. Except at the extremes, most of the data points line up fairly closely with the line of equality. Data points representing nations at the least urbanized and most urbanized ends of the scale have values higher and lower, respectively, than those predicted by a normal distribution. The term "thin tails" is sometimes used to describe this pattern. That is, there are fewer cases at the extremes than there "should" be given a normal distribution. One reason for thin tails in this case is that the range of the variable *urbpop88* is bounded by zero and 100 percent.

**Example #2:** For purpose of contrast, we'll now make a normal quantile plot of the variable *gnpcap87*, which gives each nation's GNP per capita in 1987. Is this variable normally distributed?

**1**   From the *File* menu, choose *Open.*

**2**   Enter the name of the file: humandev.

**3**   From the *Graphs* menu, choose *One variable,* and then *Normal quantile plot.*

**4**   **Windows:** Click on the data variable named *gnpcap87.*

     **DOS:** Enter the data variable named gnpcap87.

The results appear in Figure 6.10.

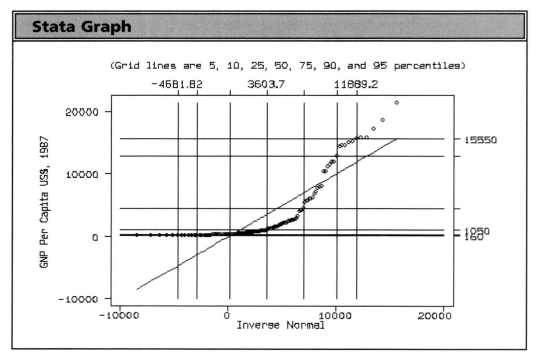

**Figure 6.10** *Normal Quantile Plot of GNP per Capita in 1987 in 130 Nations.*

A glance at Figure 6.10 reveals that the variable *gnpcap87* is not normally distributed. Careful study of the plot indicates that the lower tail of the distribution is truncated at the zero dollar floor of GNP per capita, while the upper tail is very "thick" and extended beyond the expected range. This pattern describes a severely positively skewed distribution. (Compare Figure 6.10 with the histogram shown in Figure 6.3.)

## ■ EXERCISES

**6.1** Open the *humandev.dta* file and make a continuous variable histogram of the variable *loggnp*. Choose 10 bins and normal curve overlay. Compare the graphical results with the histogram shown earlier in Figure 6.3. What conclusions do you draw regarding the effects of this "log transform" on the shape of the distribution of GNP per capita?

**6.2** Open the *hrsamp94.dta* file. This file contains various information on a sample of 108 Congressional districts, including a score devised by the

*National Journal* rating each representative's "economic liberalism" in voting on key bills. These scores are contained in a variable named *njecon*. Make continuous variable histograms of the variable *njecon* for Democratic and Republican groups (classified in the categorical variable *party*). Ask for a histogram of the total. What conclusions do you draw from a comparison of the two group histograms? How would you describe the shape of each distribution? How would you describe the shape of the histogram of the total?

**6.3**    Following up on the draft lottery example discussed in the text, open the *draft71.dta* file and repeat the same analysis on these recorded results of the 1971 draft lottery. This time ask for a histogram of the total. Do you see any evidence that the 1971 draft lottery produced nonrandom assignments of draft numbers to birthday numbers? How would you describe the shape of the histogram of the total?

**6.4**    Open the *humandev.dta* and make a normal quantile plot of the variable *literacy*. Based on the visual evidence, is this variable normally distributed? Why or why not?

# 7

# *Stem-and-Leaf Plots and Dot Plots*

## Ways to Graph and See Your Data

Like the histogram, StataQuest's stem-and-leaf plot and dot plot are graphical tools for studying one-variable distributions. The **stem-and-leaf plot** resembles a histogram turned on its side. Unlike the histogram, however, it displays the raw data in the graph itself. The **dot plot** is a type of fine-grained histogram in which each individual observation is plotted as a dot, and the dots are stacked vertically across the X-scale. The dot plot reveals less numerical detail than a stem-and-leaf plot. But it is equally good at showing the total count of observations in a distribution. And it gives a better overall picture of a distribution's shape and density. Neither plot has a scale measuring relative frequencies. Unlike other SQ graphs, stem-and-leaf plots and dot plots are displayed in the Results window rather than the Graphs window. These plots are plain text graphs that can be saved in a log file and edited.

## IN THIS CHAPTER

■ **Stem-and-leaf plots**

■ **Dot plots**

■ **Saving and printing**

## Stem-and-Leaf Plots

### What They Are and How to Interpret Them

Stem-and-leaf plots have served for years as quick, "back of the envelope" visual aids for analyzing data. Figure 7.1 shows a stem-and-leaf plot along with the raw data used to construct it.

```
Raw data for variable X                Stem-and-leaf plot for X

155, 203, 234, 319, 322,               1** | 55
339, 358, 387, 466, 612                2** | 03,34
                                       3** | 19,22,39,58,87
                                       4** | 66
                                       5** |
                                       6** | 12
```

**Figure 7.1** *Example of a Stem-and-Leaf Plot (One-Line Stems and Two-Digit Leaves)*

As shown in Figure 7.1, stem-and-leaf plots resemble histograms turned on their sides and convey similar information about the spread and shape of frequency distributions.

To construct a stem-and-leaf plot from a set of numbers, you first separate each number into a **stem** part and a **leaf** part. In the 10-based number system, each digit of a number is a counting unit in multiples of 10. In a 3-digit integer, for example, the first digit counts hundreds, the second tens, and the third ones. (If decimals are involved, the first digit to the right of the decimal counts tenths, the second hundredths, and so on.) Thus, the integer 466 is equal to the sum of four hundreds plus six tens plus six ones. Given a small set of integer data in the range 0 to 999, a reasonable design for a stem-and-leaf plot designates hundreds as the one-digit stems and tens and ones as the two-digit leaves. In this example, 4 is the stem and 66 the leaf. Another design is hundreds and tens as the two-digit stems and ones as the one-digit leaves. In this case, 46 is the stem and 6 the leaf.

A typical stem-and-leaf plot is constructed by writing down a column of stems (including those with no leaves) to the left of a vertical line. The next step is to record a leaf for each observation in a row to the right of its stem. Each leaf should have exactly the same physical width on the page. (See Exercise 7.1.) The leaves should be ordered from lowest to highest on each stem.

Studying the stem-and-leaf plot of X in Figure 7.1, we see that it is a roughly symmetrical distribution with half the values in the 300's. We observe a gap in the 500's and also note that the value of 612 might be an outlier. The plot thus provides useful qualitative information about the shape and spread of this distribution. In addition, because the stems are ordered from low to high in the column and the leaves are ordered from low to high on each row, we can easily find the median and compute other useful statistics directly from the graph. The median in this example is (322+339) / 2 or 330.5. The range is 612-155=457. You cannot extract such detailed results from a histogram.

Figure 7.2 describes some of the mental steps involved in constructing the stem-and-leaf plot shown in Figure 7.1. Note the use of double asterisks on the stems to indicate units of hundreds. Three asterisks would indicate units of thousands, and so on.

```
2. SEPARATE INTO STEMS & LEAVES

    155 = 1 hundred  + 55            3. CONSOLIDATE
    203 = 2 hundred  +  3
    234 = 2 hundred  + 34         1 hundred | +55
    319 = 3 hundred  + 19         2 hundred | +03,+34
    322 = 3 hundred  + 22         3 hundred | +19,+22,+39,+58,+87
    339 = 3 hundred  + 39   ⇒     4 hundred | +66
    358 = 3 hundred  + 58         5 hundred | (no cases)
    387 = 3 hundred  + 87         6 hundred | +12
    466 = 4 hundred  + 66
    612 = 6 hundred  + 12
                                              ⇓

                                 4. REDUCE AND GRAPH
            ⇑
                                    1** | 55
    1. SORT RAW DATA                2** | 03,34
                                    3** | 19,22,39,58,87
       155,203,234,319,            4** | 66
       322,339,358,387,            5** |
       466,612                     6** | 12
```

**Figure 7.2** *Steps for Constructing a Simple Stem-and-Leaf Plot*

## Options: Number of Lines, Digits per Leaf, Rounding, and Pruning

When you ask for a stem-and-leaf plot, SQ will look at your data and design a plot that seems to fit best given the number of observations you have, how spread out the data are, and other factors. You might not like what SQ gives you, however, and if that happens you have some options. You can change the **number of stem lines**. You can change the **number of digits** in your leaves. You can **round** your data. And you can **prune** the graph to eliminate stems that don't have leaves. Figure 7.3 shows the dialog box for choosing these options, if desired.

### Number of Stem Lines

To illustrate the first of these options, look again at the stem-and-leaf plot for X in Figure 7.1. Note that each stem has one line of data. All the 200's go in one line, all the 300's go in the next line, and so on. A lot of the cases seem to pile up on the 300's stem. If each stem had two lines of leaves, you could see how many cases fall in the low 300's and how many in the high 300's. Figure 7.4 shows what the stem-and-leaf plot in Figure 7.1 looks like with two-line stems and with the same two-digit leaves.

You could also choose five, ten, or more stem lines to achieve even finer resolution in the plot. (For one-digit leaves, the number of desired lines should divide 10 with no remainder—that is, 1, 2, 5, and 10. For two-digit leaves, the

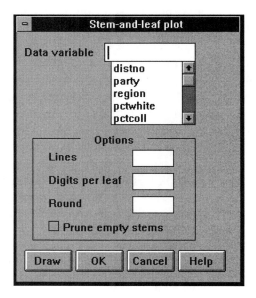

**Figure 7.3** *Stem-and-Leaf Plot Dialog Box*

```
             1** | 55          ← Leaves for X = 150 to 199 go here.
Line #1 →    2** | 03,34       ← Leaves for X = 200 to 249 go here.
Line #2 →    2** |             ← Leaves for X = 250 to 299 go here.
Line #1 →    3** | 19,22,39    ← Leaves for X = 300 to 349 go here.
Line #2 →    3** | 58,87       ← Leaves for X = 350 to 399 go here.
Etc.         4** |               Etc.
             4** | 66
             5** |
             5** |
             6** | 12
              ↑     ↑
           Stems  Leaves
          (2 lines) (2 digits)
```

**Figure 7.4** *Stem-and-Leaf Plot with Two-Line Stems and Two-Digit Leaves*

desired number should divide 100, and so on.) For the data here, one line per stem seems just right. The two lines add some resolution, but they also create more empty stems and stretch out the graph with no real gain in information.

As illustrated in Figure 7.4, the two lines for each stem have identical labels. For example, the same label 3** identifies leaves in the range 300–349 (first line) and also those in the range 350–399 (second line). This duplication of labels poses no real problem in two-line plots, but plots with five or more lines become harder to read. There is no easy solution to this problem if two or more digits per leaf are used. In two-line and five-line plots with **one** digit per leaf, however, SQ displays unique labels for each stem line.

In **two-line plots with one-digit leaves**, stems that contain the digits zero through 4 are indicated by an asterisk (*) and those that contain the digits 5 through 9 are indicated by a period (.). In **five-line plots with one-digit leaves**, stems that contain the digits zero and 1 are indicated by *; those that contain the digits 2 and 3 are indicated by t (for **t**wo and **t**hree); those that contain the digits 4 and 5 are indicated by f (for **f**our and **f**ive); those that contain the digits 6 and 7 are indicated by s (for **s**ix and **s**even); and those that contain the digits 8 and 9 are indicated by a period (.). Figure 7.5 shows two stem-and-leaf plots that illustrate these rules. Note that no commas are used to separate one-digit leaves.

### Number of Digits per Leaf

Look again at Figure 7.1. Each leaf in the plot has two digits. With only ten observations, two-digit leaves are a pretty good choice. If you have a lot of data, however, you might want to choose one-digit leaves. One digit per leaf would change stem=3** and leaf=22 to stem=32* and leaf=2. You could then place more leaves

```
Raw Data                Two-Line Plot          Five-Line Plot

    20                  2* | 03                 2* | 0
    23                  2. | 56779              2t | 3
    25                  3* | 01                 2f | 5
    26                  3. | 69                 2s | 677
    27                                          2. | 9
    27                                          3* | 01
    29                                          3t |
    30                                          3f |
    31                                          3s | 6
    36                                          3. | 9
    39
```

**Figure 7.5** *Illustration of Stem Labels in Two-Line and Five-Line Plots with One-Digit Leaves*

in a row without overflowing on the right side of the graph. One possible drawback is that the stems would be in units of ten rather than in units of one hundred. Many more stems would be needed to cover the range of the data—47 stems in all (15*, 16*, 17*, ... 59*, 60*, 61*). (Rounding is one way to deal with this problem. See next section.) With lots of data, that many stems might be needed to construct informative plots. With only ten observations, however, most of the stems would be empty and the resulting plot wouldn't tell us much.

### Rounding

Rounding offers a way to lower the number of digits per leaf without creating a lot of empty stems. SQ automatically rounds decimal tenths to the nearest integer and decimals smaller than hundredths to the nearest multiple of .01 in units of .01. It will report that it has done so with the plot. You can control the rounding by typing the desired multiple of 10 in the dialog box.

For example, to round the two-digit leaves of the plot of X in Figure 7.1 to units of ten, you would type 10 in the *Round* option box and then click on *OK*. Figure 7.6 shows the results. Note that the number 203 is rounded to 20 in units of ten. The number 387 is rounded to 39 in units of ten. Read these rounded numbers from the plot as 200 and 390. You lose some precision by rounding. But it is one way to lower the number of digits per leaf without increasing the number of stems.

### Pruning Empty Stems

Pruning is one way to eliminate empty stems. If you click this option in the dialog box and run the plot, all leafless stems will disappear from the graph.

```
ORIGINAL PLOT                      AFTER ROUNDING
 1** | 55                            1* | 6
 2** | 03,34                         2* | 03
 3** | 19,22,39,58,87               3* | 22469
 4** | 66                            4* | 7
 5** |                               5* |
 6** | 12                            6* | 1
                                   (Multiples of 10)
```

**Figure 7.6** *Stem-and-Leaf Plot Before and After Rounding*

See the illustration in Figure 7.7. Pruning empty stems is a useful option, especially if the variable being graphed has one or more severe outliers located some distance on the scale from most of the data. Use this option sparingly, however, and keep in mind that one purpose of graphing is precisely to identify gaps and outliers in a distribution.

As shown in Figure 7.7, pruning can create the illusion of no gaps or outliers. If you use such a graph in a report, be sure to indicate that it has been pruned.

```
ORIGINAL PLOT                      AFTER PRUNING
 1** | 55                            1** | 55
 2** | 03,34                         2** | 03,34
 2** |                               3** | 19,22,39
 3** | 19,22,39                      3** | 58,87
 3** | 58,87                         4** | 66
 4** |                               6** | 12
 4** | 66
 5** |
 6** | 12
```

**Figure 7.7** *Stem-and-Leaf Plot Before and After Pruning*

## A Sample Session: Stem-and-Leaf Plots

In this example we will use SQ's stem-and-leaf plot to study the distribution of energy consumption per capita in 1991 in the 50 U.S. states. This information is contained in the *energy.dta* file. The variable of interest is *btucap*, which measures energy consumption per capita in millions of British thermal units (Btu).

**1** From the *File* menu, choose *Open*.

**2** Enter the name of the file: energy.

**3**  From the *Graphs* menu, choose *One variable* and then *Stem-and-leaf plot*.

**4**  **Windows:** Click on the variable named *btucap*.

   **DOS:** Enter the variable named btucap.

The results appear in Figure 7.8.

---

**Stata Results**

```
. stem btucap

Stem-and-leaf plot for btucap (Energy Cons/Cap (mill Btu) 1991)

    1** | 97
    2** | 14,16,19,23,28,33,36,40,47
    2** | 50,85,86,91,92,92,93,93,94,94,96
    3** | 05,08,10,11,20,25,27,28,36,37,40,48
    3** | 53,67,74,80,89,92,97
    4** | 05,17,24,32,35
    4** |
    5** | 05
    5** | 64
    6** |
    6** |
    7** |
    7** |
    8** | 17
    8** | 54
    9** |
    9** |
   10** | 34
```

---

**Figure 7.8** *Stem-and-Leaf Plot of Energy Consumption per Capita (million Btu) in 1991 in the 50 U.S. States (No Options Selected)*

In Figure 7.8, we specified no options in the dialog box, so this stem-and-leaf plot is the one that SQ thinks best fits the data. It has two stem lines and two digits per leaf.

Studying the plot, we see that the distribution of *btucap* is somewhat positively skewed. All but six states fall between 200 and 449 million Btu per capita. Five states straggle out toward the high end of the scale. There is an obvious severe outlier (it turns out to be Alaska) with an energy consumption rate of 1,034 million Btu per capita. Because the stems and leaves are ordered from low to high, we can easily find the median value. Count down to the states ranked 25th and 26th, add their values, and divide by two. As you can check, the median is 315.5 million Btu. The range is 1034 minus 197 = 837 million Btu. Overall, this stem-and-leaf plot yields much that is interesting about energy consumption in the U.S.

One problem with this plot, however, is that it has quite a few empty stems. It also takes up a lot of space. We could prune the empty stems and thus produce a more compact graph, but let's not do that. (Do this on your own as an exercise.) Instead, we'll use the other options to ask for one-line stems and two digits per leaf and then run the plot again.

**1**  From the *File* menu, choose *Open*.

**2**  Enter the name of the file: energy.

**3**  From the *Graphs* menu, choose *One variable* and then *Stem-and-leaf plot*.

**4**  **Windows:** Click on the variable named *btucap* and then enter 1 in the *Lines* option box and 2 in the *Digits per leaf* box.

**DOS:** Enter the variable named btucap, 1 for *Lines*, 2 for *Digits per leaf*.

The results appear in Figure 7.9.

We see that the leaves for the 2** and 3** stems almost overflow the right side of the graph, and that is probably why SQ decided to use two-lines in the first plot. On the other hand, this plot displays essentially the same

---

**Stata Results**

```
. stem btucap, digits(2) lines(1)

Stem-and-leaf plot for btucap (Energy Cons/Cap (mill Btu) 1991)

   1** | 97
   2** | 14,16,19,23,28,33,36,40,47,50,85,86,91,92,92,93,93,94,94,96
   3** | 05,08,10,11,20,25,27,28,36,37,40,48,53,67,74,80,89,92,97
   4** | 05,17,24,32,35
   5** | 05,64
   6** |
   7** |
   8** | 17,54
   9** |
  10** | 34
```

---

**Figure 7.9** *Stem-and-Leaf Plot of Energy Consumption per Capita (millions Btu) in 1991 in the 50 U.S. States (with One Line and Two Digits per Leaf Options Selected)*

---

■ **TIP 7** ■ ■ ■ ■ ■ ■ ■ ■ ■ ■ ■ ■ ■ ■ ■

## Handling Overflows in Stem-and-Leaf Plots

Overflow occurs when a stem has too many leaves to display all of them in the plot. When that happens, SQ will show as many leaves as it can and then add a note giving the total number of leaves for that stem, including those displayed. For example, one row of the output might look like this:

```
3** | 00,13,24,32,34, [output omitted] ,55,57,61,...(53)
```

The "(53)" at the end tells you that there are a total of 53 leaves for that stem and that only some of them could be displayed. Possible solutions include asking for more stem lines, reducing the number of digits per leaf, or rounding to a higher multiple of ten.

---

information, has fewer empty stems, and is more compact. We might overrule SQ's judgment on this one.

## Dot Plots

The dot plot is a type of histogram in which each observation in the data is represented by a dot. If there are 10 observations, there will be 10 dots; 200 observations, 200 dots. For a variable X, the dots representing observations that fall into the same value range on the X scale will be vertically stacked. By comparing the relative heights of the stacked dots, you can learn something about the shape, spread, and other features of the distribution. To illustrate, let's use StataQuest's dot plot to graph the age distribution of a national sample of 596 U.S. adults (in *vote.dta*, on disk).

**1**   From the *File* menu, choose *Open*.
**2**   Enter the name of the file: vote.
**3**   From the *Graphs* menu, choose *One variable* and then *Dotplot*.
**4**   **Windows:** Click on the variable named *age*.
       **DOS:** Enter the variable named age.

The results appear in Figure 7.10.

The dot plot in Figure 7.10 lives up to its name—there are a lot of dots, 596 to be exact. As an exercise, you could count them, but we don't recommend that action. Studying this graph, we can see that the overall distribution is asym-

```
 Stata Results

. dplot age

                      . :
                     :: :
                     :: :
                    ::::.:.
               :  ::::::::              :
               :  ::::::::              :
               :  :::::::  :       :
               :  ::::::: :::   :  .
              :::::::::::::::  :  :   :
              :::::::::::::::  :     :
              :::::::::::::::  :     :
            ..:::::::::::::::.:  :       .
             :::::::::::::::::::  :   : : :.
             :::::::::::::::::::  : .:  :  : ::
           :::::::::::::::::::::.: :::::::
           :::::::::::::::::::::::::::::::::::
           ::::::::::::::::::::::::::::::::::::.   .
           :::::::::::::::::::::::::::::::::::::.     :
           ::::::::::::::::::::::::::::::::::::::::.::
      -+---------+---------+---------+---------+---------+   (596 obs.)
       0        20        40        60        80       100
```

**Figure 7.10** *Dot Plot of Age in Years for a National Sample of 596 U.S. Adults*

metrical and positively skewed. (Can you guess why that would be? Hint: The data are for **adults** 18 years or older. In technical terms, data for younger individuals has been "censored," thus chopping off the lower part of the entire age distribution.) The distribution also appears somewhat bimodal, with the larger mode in the 30–39 age group and a smaller one in the 66–79 age group. There are also some spikes that stick out like needles; these might disappear with smoothing. Because of these features, this is an interesting and informative graph.

Notice in Figure 7.10 that there is no left-side scale measuring relative frequencies. If you want to know the fraction of total observations falling in the 18–19 year old group, for example, SQ's histogram would be a better graph to use. Notice also that SQ placed tick marks at 0, 20, 40, 60, 80, and 100 years on the age scale and used the maximum of 50 bins to cover that range. The resolution achieved is an interval of two years for counting purposes, starting at 0–1 years, 2–3 years, and so on, all the way up to 99–100 years. Unfortunately, the youngest in this sample is 18 and the oldest is 89. Thus, the scale is more spread out than it needs to be to cover the raw data.

## Using the Raw Data Range

In Figure 7.10, if we could restrict the scale to a range of 18 to 89 and still use the maximum 50 bins, the age intervals would be smaller and the resolution greater. SQ's dot plot allows just that option. Rather than accept SQ's default scaling, you can choose to use the raw data to set the range. Here is how you do it.

**1**  From the *File* menu, choose *Open*.

**2**  Enter the name of the file: vote.

**3**  From the *Graphs* menu, choose *One variable* and then *Dot plot*.

**4**  **Windows:** Click on the variable named *age* and then click on the option *Use raw data range*.

   **DOS:** Enter the variable named age and answer yes to the *Use raw data range?* option.

The results appear in Figure 7.11.

## Stata Results

```
. dplot age, raw

              :
              :   :
              :   :
              :   :
            : :   :
            : :   : :
            : :   : :              .
            : :   : :     : : :
       :  : :.::  : :     : : :
       :  ::::::: : :     : : :
       :  .:::::: : :     : : :        :
       :.:::::::: :.:   : :::     :       :
      . ::::::::: ::: :::::::   : .   :    :   : : .
       : ::::::::::.:::::::::::. :.:   :    :   : : :
      :::.:::::::::::::::::::::::::: .:   .:   : : :
      ::::::::::::::::::::::::::::::::: .::   : : :
      :::::::::::::::::::::::::::::::::::.:::::::.:::: .     :
      ::::::::::::::::::::::::::::::::::::::::::::::::: :..:
     -+---------+---------+---------+---------+---------+-   (596 obs.)
     18        32.2      46.4      60.6      74.8      89
```

**Figure 7.11** *Dot Plot of Age in Years for a National Sample of 596 U.S. Adults (Raw Data Range Option Selected)*

As you can see in Figure 7.11, the age scale now ranges from 18 to 89. Instead of tick marks every 20 years, they now appear every 14.2 years and yet contain the same number of age intervals. The resolution is finer, and one result is greater detail and a general smoothing of the original plot.

### Dot Plots by Group

You can use SQ's *Graphs–One variable by group–Dot plot by group* to display two or more dot plots in one image, one for each value of a categorical variable. To illustrate, let's use this procedure to compare the age distributions of whites and nonwhites in the sample. In the *vote.dta* file, the categorical variable *race* is coded 0 for nonwhites and 1 for whites. Here are the steps:

1   From the *File* menu, choose *Open.*
2   Enter the name of the file: vote.
3   From the *Graphs* menu, choose *One variable by group* and then *Dotplot by group.*
4   **Windows:** Click on the variable named *age* as the data variable and then click on the variable named *race* as the group variable.

     **DOS:** Enter the variable named age as the data variable and race as the group variable.

The results appear in Figure 7.12.

In Figure 7.12, the dot plot at the top is for nonwhites (code 0) and the one at the bottom is for whites (code 1). To allow a meaningful comparison, the age scale is the same for both plots. The most obvious difference between the two plots is that the bottom one for whites has many more dots. The reason is that there are more whites than nonwhites in the sample–396 more, to be exact. In the dot plot by group, actual counts of cases at each value of X are compared, not the fractions of cases. Use the histogram by group (see Exercise 7.4) if you want to eliminate the distorting effects of unequal sample sizes on visual comparisons of **relative** frequencies.

## Saving and Printing

A distinctive feature of StataQuest's stem-and-leaf plots and dot plots is that both are displayed in SQ's **Results** window rather than in the **Graph** window. Stem-and-leaf plots and dot plots are **plain text files** that can be imported into any standard editor or word processor. To save and print these plots, you need to open a **log file** (see Chapter 1) **before** you run either of these procedures.

## Stata Results

```
. dplot age, by(race)

-> race=0
                     :   :      .   :
                   : :.:.:     : . :
                   :: :::::::: :::: :  :
                 .:.::::::::::: :::::.:: : .:. ...     ..
     -+---------+---------+---------+---------+---------+-   (100 obs.)
      0        20        40        60        80       100

-> race=1
                   ::
                   ::
                   :: :.
                   ::::: :   .
                 : :::::: :   :
                 : :::::: :   :
                 ::::::: ::  :
                 :::::::::::. :     :
                  :::::::::::::: . :     .
                 .:::::::::::::::: :     : .
                 :::::::::::::::::: :   : : ::
                 .:::::::::::::::::::.: ::.:.::
                 :::::::::::::::::::::::::::::::
                 :::::::::::::::::::::::::::::::. :
                 ::::::::::::::::::::::::::::::::...:
     -+---------+---------+---------+---------+---------+-   (496 obs.)
      0        20        40        60        80       100
```

**Figure 7.12** *Dot Plots of Age Distribution by Race*

The log file will capture the results for later editing and printing. In **Windows**, you can also (1) click the **Log** button to bring your active log file to the top, (2) use the mouse to select the plot text, and (3) use *Copy text* from the *Edit* menu to copy the plot to the clipboard for direct import into a document file.

## ▨ EXERCISES

**7.1**  In the stem-and-leaf plot shown in Figure 7.1, why is the two-digit "03" used rather than the simpler one-digit "3" in the 2** stem?

**7.2**  Open the *hrsamp94.dta* file (on disk). This file contains various data for a sample of 108 U.S. Congressional Districts.

 a.  Open the spreadsheet and look closely at the data for the variable *expwin*. This variable gives the campaign spending in thousands of dollars by the winning candidate in the 1994 general election. Notice the decimal.

 b.  Run a stem-and-leaf plot on the variable *expwin*. Look at the plot. What happened to the decimal?

 c.  How many lines are used in this plot and how many digits per leaf?

 d.  Describe what happens to the plot when you select the one digit per leaf option and make no other changes.

 e.  Run the plot again, this time pruning empty stems. In what way would this plot be misleading if you failed to report that it was pruned?

**7.3**  In the text discussion of Figure 7.10, it was noted that the dot plot shown there appeared bimodal. To explore one possible explanation for that bimodality, open the *vote.dta* file (on disk) and run dot plots of the data variable *age* by the group variable *sex*. Conclusions?

**7.4**  Using the *vote.dta* file (on disk), run continuous variable histograms of *age* by the group variable *race*. In the dialog box, enter 37 as the desired number of bins. Compare the resulting histograms for whites and nonwhites with the dot plots shown in Figure 7.12. Would you guess from looking at the histograms alone that whites outnumbered nonwhites in the sample by more than 4 to 1? What advantages does the histogram have over the dot plot for purposes of comparison? (Note: If you'd like actual counts rather than fractions in the histogram, run your histogram from the menu and then press the PGUP key to retrieve the last command in the Command window. Then add `freq` as an option at the end of the command and press the ENTER key.)

# Box Plots and Box and One-Way Plots

## Graphs for Seeing Both Forest and Trees

The **box plot** is a graphical tool used for analyzing one-variable distributions. It visually summarizes information about a distribution's center, spread, shape, and outliers. Box plots are particularly useful for comparing distributions of different variables or of the same variable by groups. To make a box plot of a variable X, you need a measurement scale for X, a rectangular box, a median line to put in the box, two straight lines ("whiskers") to stick in the box at either end, and some plotting symbols to identify outliers. **The one-way plot** also is a tool for univariate analysis. Its design is simple: a single measurement scale and tick marks locating each observation on the scale. It is particularly useful for studying a distribution's spread, density, gaps, and extreme values. The **box and one-way plot** combines the virtues of both types of graphs. The box plot summarizes the data; the one-way plot shows the details. One shows the forest, the other the trees.

---

▨ ▨ ▨ ▨ ▨ ▨ ▨          ■ **Box plots**

# IN THIS CHAPTER      ■ **Box plot and one-way plot**

                        ■ **Box plot by group**

                        ■ **Box plot comparisons of different variables**

                        ■ **Box plot and one-way plot comparisons**

---

# Box Plots

The best way to describe a box plot is to show you one. Figure 8.1 is an annotated StataQuest box plot of 1987 Gross National Product (GNP) in $ US per capita for 119 world nations. The name of the variable is *gnpcap87*, and it is found in the *humandev.dta* file.

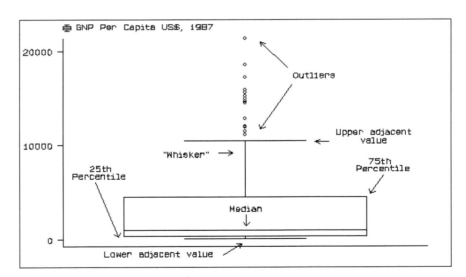

**Figure 8.1** *The Graphical Anatomy of a Box Plot*

## IQRs, Whiskers, Adjacent Values, Fences, and Outliers

The floor and ceiling of the box locate the variable's 25th and 75th percentiles, respectively, on the Y axis. The height of the box is equal to the **IQR** (interquartile range, sometimes called the midspread), computed by subtracting the

25th percentile from the 75th percentile. The horizontal line inside the box identifies the median (50th percentile). The long **whisker** poking out from the top of the box stops at the upper adjacent value. The short whisker at the bottom stops at the lower adjacent value. **Adjacent values** are the values for observations in the data that come closest to, but do not cross, the invisible **fences** (formula-defined thresholds) that separate **outliers** (extreme values) at both ends of the distribution from the rest of the data. Each outlier is identified by a plotting symbol.

## Identifying Outliers

The **lower fence** for identifying outliers is equal to the 25th percentile minus 1.5 times the IQR. The **upper fence** is equal to the 75th percentile plus 1.5 times the IQR. Using this last formula on the data for *gnpcap87*, we compute the upper fence as $4,550 + (1.5*$4,190) = $10,835. Therefore, all nations with a GNP per capita larger than $10,835 are defined as high outliers. There are sixteen high outliers in all, and each is plotted in the graph. A listing of the data shows that the United Kingdom's GNP per capita of $10,420 is the upper adjacent value and also the end-point of the top whisker. The box plot shows no low outliers.

## Interpreting the Box Plot

The box plot in Figure 8.1 tells us a lot about the distribution of GNP per capita. In 1987, half of the world's nations fell below $1,180 in GNP per capita, and one-fourth of them below $360. Half the nations had GNP per capitas greater than $1,180 and one-fourth more than $4,550. Although the range in GNP per capita was quite large, the spread of $4,190 within the middle 50 percent of nations was relatively small. Finally, the overall distribution of GNP per capita was severely positively skewed. This last conclusion is based on the low placement of the median line in the box, on the relative lengths of the whiskers, and on the presence of many high outliers combined with the absence of low outliers.

## How to Make Box Plots with StataQuest: An Example

To illustrate SQ's box plot procedure, we'll continue our study of the world's nations by asking for a box plot of average life expectancy in years in 1987. In the *humandev.dta* file (on disk), the name of this variable is *life*.

1  From the *File* menu, choose *Open*.
2  Enter the name of the file: humandev.
3  From the *Graphs* menu, choose *One variable*, and then *Box plot*.
4  **Windows:** Click on the variable named *life*.
   **DOS:** Enter the variable named life.

The results appear in Figure 8.2.

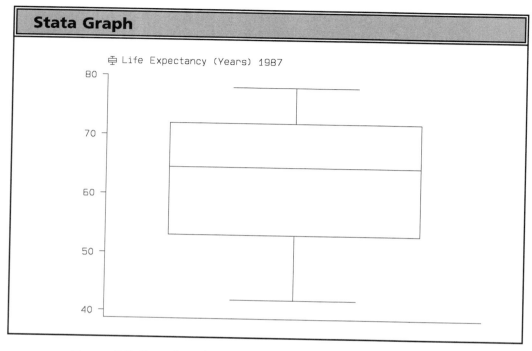

**Stata Graph**

Life Expectancy (Years) 1987

**Figure 8.2** *Box Plot of Average Life Expectancy (Years) in 1987 in the World's Nations*

The box plot in Figure 8.2 reveals that (1) the median average life expectancy of national populations is about 65 years; (2) average life expectancy ranges from a low of just over 40 years to a high of just under 80 years; (3) national populations in the middle 50 percent of the distribution have an average life expectancy of somewhere between 53 and 73 years; (4) the distribution of average life expectancy is slightly negatively skewed, and (5) there are no outliers.

## Box Plot and One-Way Plot

One limitation of the box plot is that it doesn't reveal much detail about how individual observations are distributed on the scale of measurement. The box plot in Figure 8.2, for example, is efficiently sparse. Only five numbers are plotted, and of these just two might be actual data (the lower and upper adjacent values). Within the box, it is possible that most of the observations are bunched up around the median; or perhaps they are spread out evenly across the range of the IQR. The point is that we can't find this out using the box plot alone. It is a good summary plot, but it would be nice to have more detail. SQ's *Box plot & one-way* graph answers this need.

**Example:** To illustrate the *Box plot & one-way* graph, we'll use it to re-examine the distribution of average life expectancy.

**1** From the *File* menu, choose *Open*.

**2** Enter the name of the file: humandev.

**3** From the *Graphs* menu, choose *One variable*, and then *Box plot & one-way*.

**4** **Windows:** Click on the variable named *life*.

   **DOS:** Enter the variable named life.

The results appear in Figure 8.3.

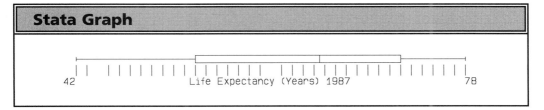

**Figure 8.3** *Box Plot & One-Way Plot of Average Life Expectancy*

Figure 8.3 combines a box plot of *life* with a one-way plot of *life*. The first summarizes the distribution, the second shows how individual nations are scattered on the same scale. Notice that the box plot is placed in a horizontal position and that the range of the scale is set by the lowest and highest values in the data (42 years and 78 years). On the accompanying one-way plot, there is a tick mark for each unique value of *life* in the data. A problem in this case is that many nations have the same value and yet are represented in the plot by only one tick mark. For example, a listing of the data shows that Ethiopia, Sierra Leone, and Afghanistan all have the same value (42 years) and yet have only one tick mark to represent them. (See Tip #9 on the use of jittering as one way to solve this problem.) Nevertheless, Figure 8.3 is still informative. Nations appear to be fairly evenly distributed across the scale and within the box. There is no evidence of bunching.

# Box Plot by Group

Box plots are especially useful for comparing distributions. The *Box plot by group* menu command allows you to compare box plots of the same variable for two or more groups defined by a second **categorical** variable.

**Example:** In 1994, the major party candidates for the U.S. House of Representatives continued their long upward trend by spending more money on

> ▨ **TIP 8** ▨ ▨ ▨ ▨ ▨ ▨ ▨ ▨ ▨ ▨ ▨ ▨ ▨ ▨ ▨
>
> ## Using String Variables to Identify Outliers in Box Plots
>
> Use the *s([string variable name])* option in the *graph [varname], box* command to identify outliers by name. For example, to produce the box plot of *gnpcap87* shown in Figure 8.1 with outliers identified by a variable called *name*, in the Command window type: `graph gnpcap87, box ylab s([name])` and press the ENTER key.

> ▨ **TIP 9** ▨ ▨ ▨ ▨ ▨ ▨ ▨ ▨ ▨ ▨ ▨ ▨ ▨ ▨
>
> ## Use Jittering to See More Detail in One-Way Plots
>
> In SQ's command mode you can produce a one-way plot by itself without the box plot. To get just a one-way plot of the variable *life*, for example, in the Command window, type: `graph life,oneway`. The *oneway* command also has a useful jittering option that will allow you to see each individual observation on the scale. Jittering slightly and randomly displaces each observation's true value on the scale to prevent ties and thus force SQ to plot them separately in the graph so they can be seen. Example: `graph life,oneway jitter(2)`. The larger the number in the *jitter(#)* option, the greater the displacement. This graph option is also available for the scatter plot. If your plot has been jittered, be sure to report that when presenting results.

their election campaigns than ever before. Using data for a sample of 108 winners in the *hrsamp94.dta* file (on disk), we'll compare box plots of campaign expenditures (in thousands of dollars) by Democrats and Republicans. The dependent variable is *expwin* and the categorical independent variable is *party*.

1   From the *File* menu, choose *Open*.
2   Enter the name of the file: `hrsamp94`.
3   From the *Graphs* menu, choose *One variable by group*, and then *Box plots by group*.
4   **Windows:** Click on *expwin* as the data variable and *party* as the group variable.

   **DOS:** Enter `expwin` as the data variable and `party` as the group variable.

The results appear in Figure 8.4.

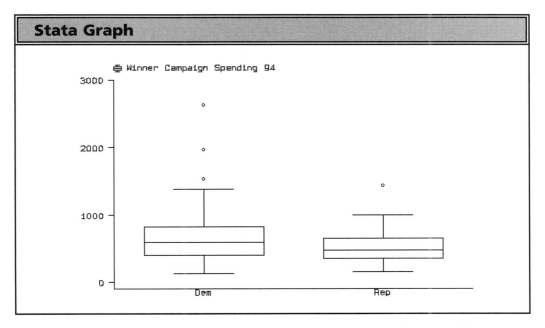

**Figure 8.4** *Box Plots by Group: 1994 Campaign Spending ($000) by Democratic and Republican Winners in Races for Seats in the U.S. House of Representatives*

Comparing the two box plots in Figure 8.4, we can see that (1) Democrats as a group spent more on their winning campaigns than did Republicans (median spending was higher); (2) there was greater variation in spending levels among Democrats than among Republicans (the IQR was larger); and (3) three Democrats and one Republican spent a lot more than their partisan colleagues in getting elected—so much more they are identified as outliers in the plot. (It would be interesting to identify these outliers by name on the plot itself. See Tip 8 and Exercise 8.2.)

*Box Plot of Total* **Option:** You might find it informative to include a box plot of the total along with the group box plots. If so, click on the *Include box plot of total* option in the dialog box.

## Box Plot Comparisons of Different Variables

Box plots can also be used to compare the distributions of different variables. We'll show two examples. The first illustrates a meaningful and informative box plot comparison. The second shows what happens when two or more of the variables being compared have dramatically different scales.

**Example #1:** The claim is made that the world's population has become more urbanized since 1960. We'll look at the evidence on that question with box plots using cross-national data on urbanization in 1960 and 1988 contained in the *humandev.dta* file. The two variables of interest are *urbpop60* and *urbpop88*. The first is the percentage of a nation's population living in urban areas in 1960; the second is the percentage living in such areas in 1988.

**1** From the *File* menu, choose *Open*.

**2** Enter the name of the file: `humandev`.

**3** From the *Graphs* menu, choose *Comparison of variables*, and then *Box plot comparison*.

**4** **Windows:** Click on the variables named *urbpop60* and *urbpop88* as the two data variables to be compared.

    **DOS:** Enter the variables named `urbpop60` and `urbpop88` as the data variables.

The results appear in Figure 8.5.

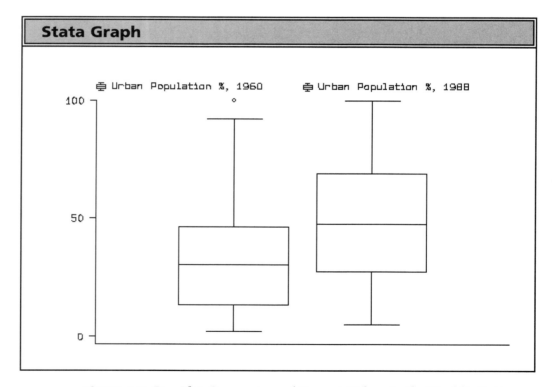

**Figure 8.5** *Box Plot Comparison of Percent Urban in the World's Nations: 1960 and 1988*

The box plot comparison in Figure 8.5 shows clearly that the world's populations were more urbanized in 1988 than in 1960. The median percentage urban rose from about 30% in 1960 to nearly 50% in 1988. The overall distribution of the percentage urban shifted upwards over this period. Indeed, the single outlier in 1960 (Singapore, 100% urban) was no longer an extreme value in 1988. Note that the box for 1988 is taller than the one for 1960, however, suggesting greater variation in urbanization among nations in the middle 50 percent.

The box plot comparison displayed in Figure 8.5 is meaningful and informative because both variables are measured on the same scale (0 to 100%). The next example illustrates what happens when the variables being compared are not measured on the same scale.

**Example #2:** Using the *humandev.dta* file, let's do a box plot comparison of GNP per capita in 1987 and percentage urban in 1988. The variables to be compared are *gnpcap87* and *urbpop88*.

**1** From the *File* menu, choose *Open*.

**2** Enter the name of the file: `humandev`.

**3** From the *Graphs* menu, choose *Comparison of variables*, and then *Box plot comparison*.

**4** **Windows:** Click on the variables named *gnpcap87* and *urbpop88* as the two data variables to be compared.

  **DOS:** Enter the variables named `gnpcap87` and `urbpop88` as the data variables.

The results appear in Figure 8.6.

Figure 8.6 shows that the distributional characteristics of *gnpcap87* can be clearly read from the box plot for that variable. The box plot for *urbpop88*, however, is squashed into a thin pancake that tells us little about that variable. No meaningful comparison of the two box plots is possible. The Y-axis scale had to accommodate the range of values in both variables, which worked just fine for *gnpcap87*. The unit of measure on the Y-axis can be safely interpreted as dollars per capita ranging from zero to more than 20,000. The *urbpop88* variable ranges only from zero to 100, however, so the box plot had to be compressed within a tiny interval on the single scale. One way around this problem is to use SQ's *Box plot & one-way comparison* as an alternative.

## Box Plot and One-Way Plot Comparisons

This graphical tool provides a way to make box plot comparisons of different variables even if the variables are measured on dramatically different scales. When using this comparison plot, you have the option to "scale each

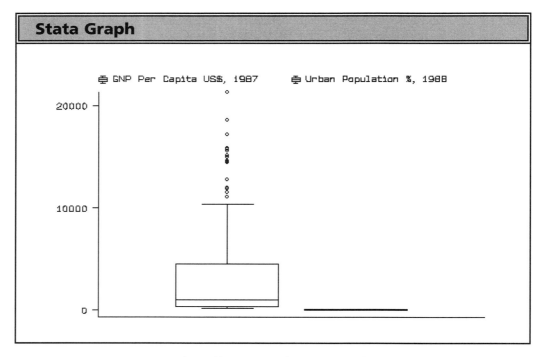

**Figure 8.6** *Box Plots Illustrating the Mistake of Comparing Variables that have Dramatically Different Scales*

variable." If you check that option, each variable shown in the graph will be plotted on its **own** measurement scale. Our first example illustrates the output obtained without using the scaling option. The second shows the effects of scaling to each variable.

**Example #1 (Without Scaling):** Using the *humandev.dta* file, let's do a box plot + one-way plot comparison of GNP per capita in 1987 and percentage urban in 1988. The variables to be compared are *gnpcap87* and *urbpop88*.

**1**  From the *File* menu, choose *Open*.

**2**  Enter the name of the file: humandev.

**3**  From the *Graphs* menu, choose *Comparison of variables*, and then *Box plot & one-way comparison*.

**4**  **Windows:** Click on the variables named *gnpcap87* and *urbpop88* as the two data variables to be compared.

   **DOS:** Enter the variables named gnpcap87 and urbpop88 as the data variables.

The results appear in Figure 8.7.

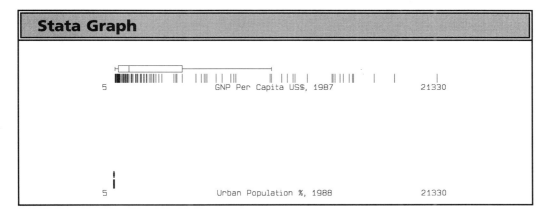

**Figure 8.7** *Box Plot and One-Way Plot Comparison of GNP per Capita and Percent Urban in the World's Nations (without Scaling Each Variable)*

Figure 8.7 shows that this comparison of the two variables does not improve much on the output obtained earlier using *Box plot comparisons* (see Figure 8.6). The one-way plot for *gnpcap87* reveals more detail about that variable's distribution. But the box plot and one-way plot for *urbpop88* are pancaked as before and are impossible to read.

**Example #2 (With Scaling):** Using the *humandev.dta* file (on disk), we'll do another box plot + one-way plot comparison of GNP per capita in 1987 and percentage urban in 1988. The variables to be compared are *gnpcap87* and *urbpop88*. This time we'll use the "scale each variable" option.

**1** From the *File* menu, choose *Open*.

**2** Enter the name of the file: humandev.

**3** From the *Graphs* menu, choose *Comparison of variables*, and then *Box plot & one-way comparison*.

**4** **Windows:** Click on the variables named *gnpcap87* and *urbpop88* as the two data variables to be compared, and then click on the *Scale each variable* option.

   **DOS:** Enter the variables named gnpcap87 and urbpop88 as the data variables and accept the *Scale each variable* option.

The results appear in Figure 8.8.

The box plots and one-way plots shown in Figure 8.8 can be meaningfully compared. For example, we can see that the distribution of GNP per capita is severely positively skewed and has many high outliers. The distribution of percentage urban, on the other hand, is approximately symmetrical and has no outliers.

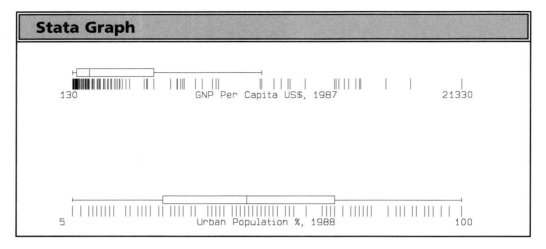

**Figure 8.8** *Box Plot and One-Way Plot Comparison of GNP per Capita and Percent Urban in the World's Nations (with Scaling Each Variable)*

## ■ EXERCISES

**8.1**   Open the *xyzbox.dta* file (on disk) and use SQ's *Comparison of variables–Boxplot comparison* to display box plots of the variables named *x*, *y*, and *z*. What conclusions do you draw from this comparison? Now do a stem-and-leaf plot or a dot plot of each of these three variables. What conclusions do you draw? What does this exercise suggest regarding the limitations of box plots as graphical tools?

**8.2**   In the box plots in Figure 8.4, three outliers are shown for Democrats and one for Republicans. Use the steps suggested in Tip 8 to identify these outliers by name on the plot itself. (*Hint:* The data file includes a string variable, *name*, which gives the last name of the winning candidate.)

**8.3**   Open the *hrsamp94.dta* file (on disk) and run SQ's *One variable–Box plot & one-way* on the variable *party*. What do the results tell you about the types of variables most appropriate to use with these graphical tools?

**8.4**   Open the *vote.dta* file (on disk) and run SQ's *One variable by group–Boxplots by group* on the data variable *age* and the group variable *sex*. Is there evidence of a gender difference in age distributions? Explain.

**8.5**   Open the *energy.dta* file (on disk) and use box plots to compare the four major U.S. regions in energy consumption per capita. Based on a comparison of medians, which region has the highest energy consumption per capita, and which region has the lowest? Which region has the most variation among states in energy consumption per capita, and which region has the least variation? Which regions, if any, have outliers?

# Bar Charts

## How Much? Graphs for Comparing Variables and Groups

Bar charts are useful graphs for visually comparing how much one has of one thing versus how much one has of another thing—or for comparing how much one has of one thing versus how much another has of the same thing. The dashboard of a typical automobile, for example, has several animated bar charts of a sort to guide our driving decisions. A tall bar touching "Hot" on our temperature gauge combined with a short bar near "Empty" on the fuel gauge probably spells trouble. Similar bars reading "Cool" and "Full," respectively, on our neighbor's car, might cause us to regret making the comparison.

One kind of bar chart might display four bars of varying heights, a single bar for each region, to compare how many people live in each region in total (adding up the populations of each region's states) or on average (dividing each region's total population by the number of states). Such a bar chart would compare four "groups" (regions) on one "data variable" (population). Another kind of bar chart might display three bars of varying heights, the first measuring an industry's total revenues, the second its total expenditures, and the third its total profits. Such a bar chart would compare industry totals for three

---

| | | | | | | ■ **Bar charts: totals and means** |
| | | | | | | |

■ ■ ■ ■ ■ ■ ■      ■ **Bar charts: totals and means**

## IN THIS CHAPTER    ■ **Bar charts by group**

■ **Bar chart comparisons**

■ **Bar chart comparisons by group**

---

data variables on a common scale of dollars. This chapter will illustrate how to use StataQuest to produce these and other kinds of bar charts.

## Bar Charts: Totals and Means

StataQuest's bar chart will display bar heights proportional to the **total** or to the **mean** of each variable specified. SQ assumes you want totals. If you click on the *Bar height to reflect means* option in the dialog box, SQ will give you means.

To illustrate the difference in output, let's produce bar charts of the data shown in Figure 9.1 on industrial-use energy consumption (trillions of Btu) by state and by region. The data are contained in the *energy.dta* file.

The entries in the bottom three rows of Figure 9.1 report the total energy consumption for all states in each region (Totals), the number of states in each region (Ns), and the mean energy consumption in each region (Means). These statistics were computed using SQ's *Summaries–Means and SDs by group* command. In Figure 9.2, the bar heights in the chart on the left are proportional to the totals for each region. The bar heights in the chart on the right are proportional to the means for each region.

Both bar charts in Figure 9.2 show that in 1991 the South was the heaviest user of energy for industrial purposes. Notice, however, that the regional differences are visibly less dramatic in the bar chart of means than in the bar chart of totals. The South includes sixteen states, more than in any other region. That fact alone could explain why the south as a whole consumed the most total energy. The bar chart of means standardizes the regional comparison by focusing on average energy consumption **per state** within each region.

## Bar Charts by Group

SQ's *Bar charts by group* menu command produces a bar chart of totals or means for one **data variable**. The chart will display one bar for each value of a categorical **group variable**.

| | NORTHEAST | | MIDWEST | | SOUTH | | WEST |
|---|---|---|---|---|---|---|---|
| | CT | 135 | IA | 347 | AL | 731 | AK | 334 |
| | MA | 213 | IL | 1219 | AR | 268 | AZ | 184 |
| | ME | 119 | IN | 1147 | DE | 90 | CA | 1922 |
| | NH | 54 | KS | 422 | FL | 418 | CO | 234 |
| | NJ | 539 | MI | 927 | GA | 630 | HI | 67 |
| | NY | 693 | MN | 485 | KY | 638 | ID | 146 |
| | PA | 1302 | MO | 346 | LA | 2217 | MT | 146 |
| | RI | 50 | ND | 159 | MD | 379 | NM | 189 |
| | VT | 26 | NE | 132 | MS | 358 | NV | 106 |
| | | | OH | 1490 | NC | 632 | OR | 276 |
| | | | SD | 53 | OK | 515 | UT | 217 |
| | | | WI | 470 | SC | 495 | WA | 700 |
| | | | | | TN | 724 | WY | 236 |
| | | | | | TX | 5490 | | |
| | | | | | VA | 466 | | |
| | | | | | WV | 423 | | |
| **Totals:** | 3131 | | 7197 | | 14474 | | 4757 |
| **Ns:** | 9 | | 12 | | 16 | | 13 |
| **Means:** | 347.9 | | 599.8 | | 904.6 | | 365.9 |

**Figure 9.1** *Table of Industrial End-Use Sector Energy Consumption in Trillions of Btu, by State and by Region, 1991 (U.S.Bureau of the Census, Statistical Abstract of the United States: 1994)*

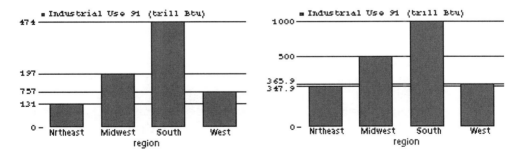

**Figure 9.2** *Bar Charts of Data in Figure 9.1: Totals by Region and Means by Region*

For example, if your data variable is campaign expenditures by candidates for the U.S. Congress and your group variable is political party, your bar chart by group will have two bars. The first will show total spending by Democrats and the second will show total spending by Republicans. In the example shown in Figure 9.2, the data variable is energy consumption and the group variable is region. Because there are four regions, the bar chart by group has four bars, one for each region.

You have two **options** in the dialog box for *Bar charts by group*: (1) Click on the *Bar height to reflect means* option to display bar heights proportional to means for each group. (2) Click on *Include bar chart of total* option to add a bar to the graph showing totals or means of the pooled data for all groups. If you choose neither option, SQ will assume you want one bar showing the total for each group specified. You can also select just one option or both if you want.

**Example:** To illustrate the *Bar chart by group* menu command, we'll open the *energy.dta* file and ask for a bar chart of the data variable *indus* by the group variable *region*.

**1** From the *File* menu, choose *Open*.

**2** Enter the name of the file: energy.

**3** From the *Graphs* menu, choose *One variable by group*, and then *Bar charts by group*.

**4** **Windows:** Click on the variable named *indus* as the data variable and then on the variable named *region* as the group variable. Click on the *Bar height to reflect means* option and then on the *Include bar chart of total* option.

**DOS:** Enter the variable named indus as the data variable and then the variable named region as the group variable. Enter yes to accept the *Bar height to reflect means*. The bar chart of total menu option is not available in the DOS version.

The results appear in Figure 9.3.

The bar chart by group shown in Figure 9.3 essentially replicates the second bar chart displayed earlier in Figure 9.2, with the addition of a new bar (labeled "Total") giving the mean energy consumption for all 50 states. Southern states as a group are clearly above the overall state average in industrial-use energy consumption. One might conclude from this chart that southern states have replaced the midwestern states as America's industrial heartland. (But see Exercise 9.5.)

## Bar Chart Comparisons

StataQuest's *Bar chart comparison* menu produces a bar chart of totals or means for **two or more data variables**. The chart will display one bar for each **variable** selected. SQ will display a different color for each bar on a color monitor and will shade the bars differently in printing the graph. Click on the *Bar height to reflect means* option in the dialog box to display bar heights

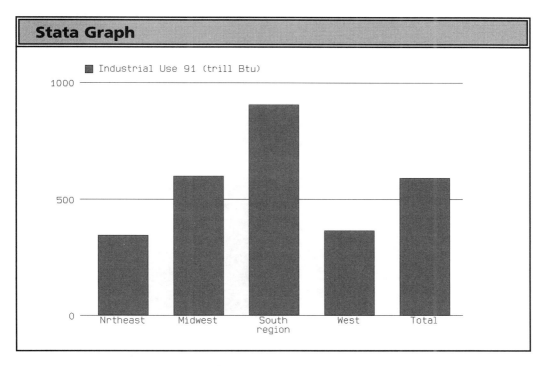

**Stata Graph**

**Figure 9.3** *Bar chart of Mean State Industrial End-Use Sector Energy Consumption in Trillions of Btu, by Region, 1991. The Mean for all 50 U.S. States Is Shown (Total)*

proportional to means for each variable. If you don't select that option, SQ will assume you want the bar heights to reflect totals for each variable. (For additional options in SQ's command mode, see the Tip boxes later in this chapter.)

**Example:** Total energy consumption is divided into four types of end-use sectors: residential, commercial, industrial, and transportation. Which of these end-use sectors consumes most of the energy produced in U.S. states? Which consumes the least? The *energy.dta* file has 1991 energy consumption data categorized by end-use sectors for each of the 50 U.S. states. The variables are named *resid, comm, indus,* and *transp.* We'll run a bar chart comparison of these four variables. This time we want the totals (in trillions of Btu) for each variable that SQ automatically provides. So we do **not** click the option box for means.

**1**  From the *File* menu, choose *Open*.

**2**  Enter the name of the file: energy.

**3**  From the *Graphs* menu, choose *Comparison of variables*, and then *Bar chart comparison*.

**4** **Windows:** Click on the variables named *resid*, *comm*, *indus*, and *transp* as the data variables. Click on the *Bar height to reflect means* option.

**DOS:** Enter the variables named `resid`, `comm`, `indus`, and `transp` as the data variables. Choose the *Bar height to reflect means* option.

The results appear in Figure 9.4.

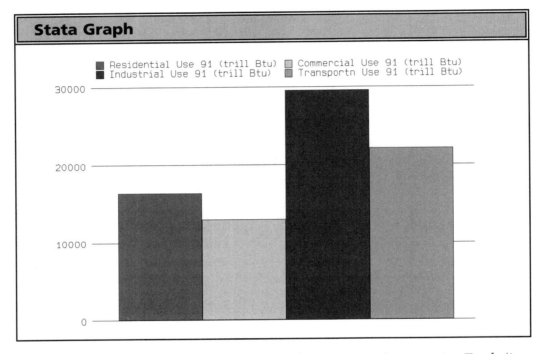

**Figure 9.4** *Bar Chart Comparison of 1991 Energy Consumption Totals (in Trillions of Btu) in the Residential, Commercial, Industrial, and Transportation End-Use Sectors: All 50 U.S. States*

Notice in Figure 9.4 that each bar has a different shading to distinguish it from the others. The legend at the top of the chart keys each shading to a sector type.

The bar chart comparison in Figure 9.4 shows clearly that the industrial sector claimed the lion's share of energy produced in 1991 (nearly 600 trillion Btu), followed by the transportation sector (more than 400 trillion Btu). The residential sector ranked third, claiming more than 300 trillion Btu to serve the housing energy needs of 260 million Americans living in 50 states. The commercial sector made the smallest aggregate demand on energy (less than 300 trillion Btu).

# Bar Chart Comparisons by Group

SQ's *Bar chart comparison by group* menu command displays a bar chart comparison **of two or more data variables** for each value of a categorical **group variable**. This type of bar chart allows three distinct kinds of visual comparison to be made in one graphical image: (1) Totals or means of a single variable can be compared across groups. (2) Totals or means of several variables can be compared within each group. (3) The overall pattern of differences among variables observed in one group can be compared with the overall pattern observed in other groups—that is, the within-group bar chart comparisons of variables can themselves be compared across groups.

As with *Bar charts by group*, you have two **options** in the dialog box for *Bar chart comparison by group*: (1) Click on the *Bar height to reflect means* option to display bar heights proportional to means for each group. (2) Click on *Include bar chart of total* option to add a bar to the graph showing totals or means of the pooled data for all groups. If you choose neither option, SQ will assume you want one bar showing the total for each specified data variable within each group. Thus, if you specify four data variables and one group variable with four categories, SQ will display sixteen bars. You can also select just one option or both if you wish.

**Example:** To illustrate the *Bar chart comparison by group* menu command, we'll open the *energy.dta* and ask for a bar chart comparison of the data variables *resid*, *comm*, *indus*, and *transp* by the group variable *region*. We'll pass on both options.

**1**  From the *File* menu, choose *Open*.

**2**  Enter the name of the file: `energy`.

**3**  From the *Graphs* menu, choose *Comparison of variables*, and then *Bar chart comparison by group*.

**4**  **Windows:** Click on the variables named *resid*, *comm*, *indus*, and *transp* as the data variables and then on the variable named *region* as the group variable.

**DOS:** Enter the variables named `resid`, `comm`, `indus`, and `transp` as the data variables and then the variable named `region` as the group variable.

The results appear in Figure 9.5.

As one approach to interpreting Figure 9.5, we'll systematically illustrate the three kinds of comparison mentioned.

**Comparing one variable across groups:** The darkest-shaded bar in each of the four groups (third bar from the left in each group) reflects the total energy consumed for industrial use within each region. This comparison of one variable across groups is essentially identical to that shown earlier in the left bar

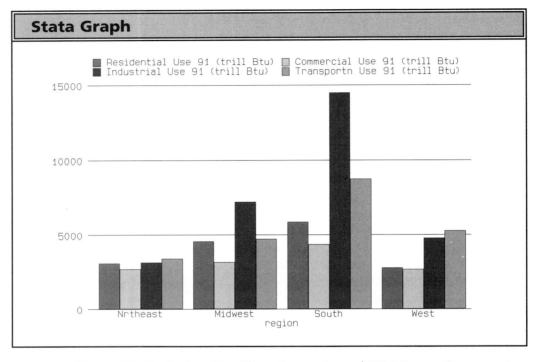

**Figure 9.5** *By Region: Bar Chart Comparison of 1991 Energy Consumption Totals (in Trillions of Btu) in the Residential, Commercial, Industrial, and Transportation End-Use Sectors*

chart in Figure 9.2. You can repeat this same kind of "by group" comparison separately for each of the other three variables.

**Comparing several variables within each group:** Notice that the Northeast group's overall energy budget is fairly evenly distributed among the four end-use sectors. You can repeat this same kind of comparison of variables separately for each of the other three regions.

**Comparing overall patterns of differences among variables across groups:** Comparing the South with the Northeast, notice that the South's overall energy budget is much less evenly distributed among the four end-use sectors. Further, the energy consumption of the South exceeds that of the Northeast (and that of all other regions, for that matter) in every one of the four end-use sectors. A careful study of this chart will reveal many other regional differences in overall patterns of energy consumption.

## ■ TIP 10 ■ ■ ■ ■ ■ ■ ■ ■ ■ ■ ■ ■ ■ ■ ■ ■

### "Stacked" Bar Charts

In SQ's command mode, you can add the "stack" option to the *graph* command to produce the stacked bar chart comparison of variables by group shown in Figure 9.6. Instead of reading each region's bars from left to right as in Figure 9.5, in Figure 9.6 you read the bars in the same order from the bottom to the top: residential, commercial, industrial, transportation. The vertical height of each layer in a stack is proportional to that region's total energy consumption in the relevant sector. One advantage of stacked bars over those shown in Figure 9.5 is that bar heights reflect regional differences in **total** energy consumed and can be more easily seen at a glance. The two SQ commands that produced Figure 9.6 are the following:

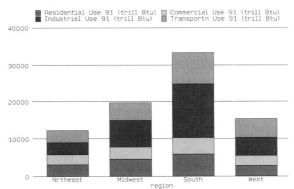

**Figure 9.6** *A "Stacked" Version of the Bar Chart Comparison by Group Shown in Figure 9.5. Produced by the "stack" option in SQ's graph command.*

```
(1) sort region
(2) graph resid comm indus transp, by(region)
    b2title(region) bar ylab noax ylin stack
```

The only difference between the second command and the one that produced Figure 9.5 is the addition of the word *stack* at the end of the command line.

Thus, easy steps for creating stacked bar charts in SQ for Windows are these:

1. Run the *Bar chart comparison by group* menu command, which will automatically sort the specified group variable.
2. Either press the PGUP key or click on the last command in the Review window to enter it in the Command window.
3. Type `stack` at the end of the command line and then press the ENTER key.

**Important Note:** SQ will **not** produce stacked bar charts if you select the *Bar height to reflect means* option in the dialog box.

▧ **TIP 11** ▨ ▨ ▨ ▨ ▨ ▨ ▨ ▨ ▨ ▨ ▨ ▨ ▨ ▨

## Creating Bar Charts for One or a Few Observations

In this chapter you have learned how to produce bar charts of a single data variable for each category of a group variable, bar chart comparisons of several data variables, and bar chart comparisons of several data variables for each category of a group variable. But suppose you wanted to create a bar chart comparison of several data variables for just one or a few observations in your data set. Using the *energy.dta* file, for example, let's say you want a bar chart comparison of the data variables *resid*, *comm*, *indus*, and *transp* for only three states (not whole regions), and all in one graph. How would you do that? Here is one way to proceed in SQ's command mode. Assume the three states of interest are California, Texas, and New York. The group variable in this case is not *region* but *state*. In SQ's Command window, enter and execute the following two commands:

```
(1) sort state
(2) graph resid comm indus transp if
    state == "California"  |  state == "Texas" |
    state =="New York", by(state)
    b2title(state) bar ylab noax ylin
```

The output is shown in Figure 9.7. For an explanation of the use of "if" statements, see Chapter 4. Try the *sort* and *graph* commands shown here to verify that they produce the same results. Try some variations to get the feel of how it works (for example, drop Texas and New York).

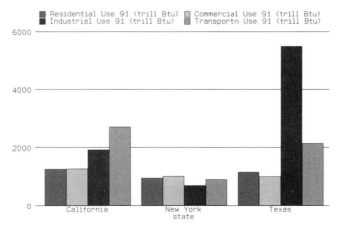

**Figure 9.7** *Bar Chart Comparisons for Three States.*

## ■ EXERCISES

**9.1** Open the *states3.dta* file and run a bar chart by group with *pop1993* as the data variable and *region* as the group variable. What conclusions do you draw from the resulting bar chart about the regional distribution of population in the United States? Run the same analysis again, this time asking for means rather than totals. Would you change or qualify your initial conclusions based on these new results?

**9.2** Open the *highway3.dta* file. This file contains various statistics on highways, traffic volume, accidents, and driving behavior for the U.S. states. Run a bar chart by group with the number of fatal vehicle accidents in 1993 (*numfatac*) as the data variable and *region* as the group variable. Then run a bar chart by group with the number of vehicle miles traveled (in millions) in 1993 (*anvehmi*) as the data variable and *region* as the group variable. What conclusions do you draw from these two charts regarding regional differences in fatal accident rates?

**9.3** Open the *hrsamp94.dta* file. This file contains various information on a sample of 108 members of the U.S. House of Representatives. Choose and run the appropriate bar chart command to provide a visual test of the following hypothesis: "Republicans, on average, spent more money winning their campaigns in the 1994 general election than did Democrats—and they did so, on average, against less well-funded challengers. Does the graphical evidence support this hypothesis? Explain.

**9.4** Many serious analysts discourage the use of bar charts such as the one displayed in Figure 9.3 because such graphs have a very low "information-to-ink ratio." Judge for yourself. Open the *energy.dta* file and use *Summaries–Means and SDs by group–One-way of means* with *indus* as the data variable and *region* as the group variable. Look at the column of five means in the resulting table. What does the bar chart in Figure 9.3 say or do that these five numbers don't say or do more precisely? Which would you prefer to include in a report, the bar chart or a table reporting the five means? Explain.

**9.5** (This exercise assumes you understand how to produce and interpret SQ's box plots. See Chapter 8.) In the text discussion of the bar chart by group in Figure 9.3, the statement was made that one "might conclude from this chart that southern states have replaced the midwestern states as America's industrial heartland." First, look at the listing of data for southern states in Figure 9.1. What effect do you think the inclusion of Texas in this list might have on the calculation of the mean industrial-use energy consumption for this region? Second, repeat the graphical analysis that produced Figure 9.3, only this time use SQ's *Box plots by group* as the tool of choice. Based on these box plot results, how would you evaluate the statement above regarding America's industrial heartland? What do you conclude from this exercise about the relative strengths and weaknesses of bar charts compared with box plots?

# 10

## Scatter Plots

### What are Scatter Plots?

Scatter plots are graphical tools used for analyzing two-variable relationships. A scatter plot is drawn with two axes. The vertical Y-axis on the left of the plot has a scale for the dependent Y variable. The horizontal X-axis on the bottom has a scale for the independent X variable. To plot data in this graph, you need a number for the Y-scale and one for the X-scale for each observation. These two numbers are called the Y,X coordinates for a particular observation. If your data set contains 100 observations, you will need 100 coordinates or pairs of Y,X numbers to draw a scatter plot. These coordinates identify the location (or "address") of each observation in the Y versus X graph space. A plotting symbol such as a dot or a circle is then used to mark the location of each observation in that space. When that is done, you have a scatter plot. With it, you can actually see the pattern of relationship that exists between Y and X.

Here is an example that plots data on income and education for the 50 U.S. states (from *states3.dta*). The Y variable is *income93*, per capita income in 1993. The X variable is *baplus*, the percentage of state residents older than 25 years who held at least a BA degree in 1990. Figure 10.1 shows a snippet of the data for five states along with the Y,X coordinates to be used in plotting.

■ ■ ■ ■ ■ ■ ■

## IN THIS CHAPTER

- **Plot Y vs. X**
- **Plot Y vs. X, naming points**
- **Plot Y vs. X, with regression line**
- **Plot Y vs. X, scale symbols to Z**
- **Plot Y. vs. X, by group**
- **Scatter plot matrix**

| state | income93 | baplus | Y,X Coordinates |
|---|---|---|---|
| Nevada | 22729 | 15.3 | (22729, 15.3) |
| New Hampshire | 22659 | 24.3 | (22659, 24.3) |
| New Jersey | 26967 | 24.8 | (26967, 24.8) |
| New Mexico | 16297 | 20.4 | (16297, 20.4) |
| New York | 24623 | 23.1 | (24623, 23.1) |

**Figure 10.1** *Listing of Data for Five States on* income93 *(Y) and* baplus *(X)*

Figure 10.2 shows a scatter plot of *income93* by *baplus* for the 50 U.S. states, including the five listed in Figure 10.1. Arrows are used to identify the symbols for Nevada and New Jersey in relation to their plotting coordinates on the Y and X scales. (Can you find New Hampshire, New Mexico, and New York?)

It is apparent from looking at this plot that (1) there is a positive relationship between income and education, (2) the relationship is fairly strong (not much "scatter"), (3) the relationship is approximately linear, and (4) some states, including Nevada, seem to stand off a bit from the main pattern. Like most scatter plots, Figure 10.2 displays valuable information about the direction (positive or negative), strength (strong or weak), and functional form (linear or nonlinear) of a two-variable relationship. It also helps to identify individual observations that are at odds with the overall pattern.

SQ's *Scatterplots* menu offers a choice of six different kinds of scatter plots (see Figure 10.3):

- *Plot Y vs. X* gives you the standard scatter plot of Y by X, with circles used as the plotting symbol.
- *Plot Y vs. X, naming points* enables you to use names and other identifiers as plotting symbols.

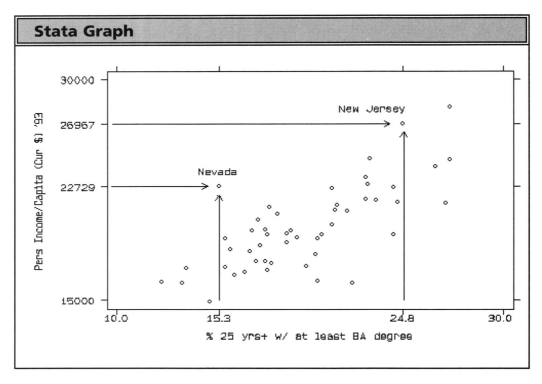

**Figure 10.2** *Scatter Plot of Income per Capita by Educational Attainment in the 50 States*

■ *Plot Y vs. X, with regression line* will overlay the standard plot with the model fit line of predicted Y values ("Y-hats") from a simple regression of Y on X.

■ *Plot Y vs. X, scale symbols to Z* will make the plotting symbol bigger or smaller in proportion to an observation's value on a third variable.

■ *Plot Y vs. X, by groups* will display in one image a scatter plot of Y vs. X for each value of a categorical variable (for example, regions).

■ *Scatter plot matrix* will display in one image all possible Y vs. X (and X vs. Y) scatter plots for lists of three or more variables.

This chapter will show the dialog box for each type of scatter plot, illustrate the typical output, and briefly discuss how to use it as a tool of analysis.

To keep things focused, we'll continue to use the *states3.dta* file and stick with *income93* as the Y variable and *baplus* as the X variable in most examples.

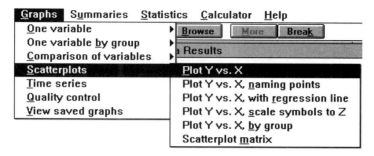

**Figure 10.3** *Menu Options for Scatterplots*

# Plot Y vs. X

This gives you the standard, no-frills, Y by X scatter plot.
**Example.** To draw a scatterplot, do the following:

1  From the *File* menu, choose *Open*.
2  Enter the name of the file: states3.
3  From the *Graphs* menu, choose *Scatterplots* and then *Plot Y vs. X.*
4  **Windows:** In the dialog box (see Figure 10.4), use the mouse to scroll the list and to click on *income93* as the Y-axis variable. Similarly, enter *baplus* as the X-axis variable. Click on Draw.

    **DOS:** In answer to the questions, the Y-axis variable should be income93 and the X-axis variable, baplus.

The result is shown in Figure 10.5.

**Note:** In Windows, click on the Draw button if you intend to run a number of Y vs. X scatter plots on this data set. That will keep the dialog box on the screen while various graphs are displayed. Otherwise, click on the OK button. The dialog box will disappear and only the graph will be shown.

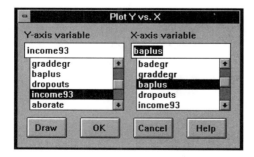

**Figure 10.4** *Dialog Box for Plot Y vs. X*

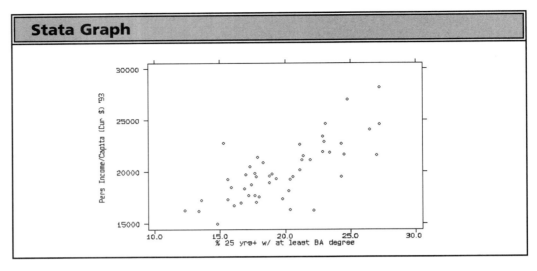

**Figure 10.5** *Output of Plot Y vs. X*

The scatter plot shown in Figure 10.5 is the same as that displayed earlier in Figure 10.2—minus the arrows and annotations.

## Plot Y vs. X, Naming Points

This plot is the scatter plot to use if you'd like to see more information about the data **in the graph**. One way to do that is to replace the circles with something more informative as plotting symbols. In the example that follows, we'll select *abvstate* as the "labeling variable" to use abbreviated names of the states as plotting symbols. But the values of any variable in the data set (to a maximum of eight characters) could also be used for this purpose. You might want to use state high school dropout rates as plotting symbols, for example, or state population size in thousands. By "naming points" in this fashion you can, in effect, add a third variable to the graphical analysis.

**Example:** The scatter plot in Figure 10.5 is interesting and tells us a lot about the relationship between *income93* and *baplus*. But we'd also like to know the identities of the states shown in the plot. The *states3.dta* file contains two string variables, *state* and *abvstate*, that we can use to label the points in the graph. We choose *abvstate* as the labeling variable because it is only two characters wide and will make the plot easier to read.

**1**   From the *File* menu, choose *Open*.

**2**   Enter the name of the file: `states3`.

**3**   From the *Graphs* menu, choose *Scatterplots* and then *Plot Y vs. X, naming points*.

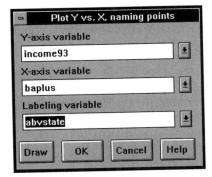

**Figure 10.6** *Dialog Box for* Plot Y vs. X naming points

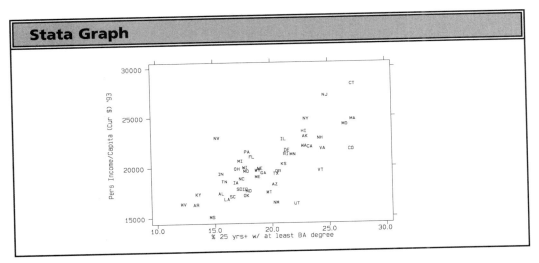

**Figure 10.7** *Output of* Plot Y vs. X, naming points

**4** **Windows:** In the dialog box (see Figure 10.6), use the mouse to scroll the list and to click on *income93* as the Y-axis variable. Similarly, enter *baplus* as the X-axis variable, and *abvstate* as the labeling variable. Click on Draw.

**DOS:** In answer to the questions, the Y-axis variable should be `income93`, the X-axis variable, `baplus`, and the labeling variable, `abvstate`.

The result of the procedure is shown in Figure 10.7.

The use of state names as plotting symbols adds a new, third dimension to the analysis. We now see that those two states in the upper-right part of the plot (high education, high income) are Connecticut and New Jersey. The three in the lower left (low education, low income) are Arkansas, Mississippi, and West Virginia. Nevada sticks out as a state with a much higher income per capita than it "should" have given its rather low level of educational attainment. Similarly, Utah, Vermont, and Colorado all seem to have lower levels of income than they "should" have given their levels of education. These preliminary "eyeball" assessments might offer clues as to next steps to take in the analysis. The "deviant" cases of Nevada and Utah are particularly intriguing. Now that we know their identities, we might want to dig a bit deeper to find out why they don't fit the overall pattern revealed by the scatter plot.

# Plot Y vs. X, with Regression Line

This command draws a regression line on the standard Y vs. X scatter plot and prints the regression equation at the top of the graph. The output is essentially identical to that of the *Plot fitted model* diagnostic graph following simple regression (see Chapter 17). In this case, however, you don't have to run the regression first. SQ does that for you, quietly (quietly means no output appears on the screen) and with no statistics, and uses only the regression line to show you how well a linear model fits the observed pattern of relationship between Y and X.

   **Example.** The relationship between *income93* and *baplus* looks approximately linear. But our eyes can deceive us. Let's take a quick look by plotting *income93* vs. *baplus* with a regression line. If the regression line fits the data pattern reasonably well, we can use the regression equation that produced that line as a compact description of the relationship between these two variables. If it doesn't fit very well, we still learn something: the relationship isn't linear.

**1**   From the *File* menu, choose *Open*.
**2**   Enter the name of the file: `states3`.
**3**   From the *Graphs* menu, choose *Scatterplots* and then *Plot Y vs. X, with regression line*.
**4**   **Windows:** In the dialog box (see Figure 10.8), use the mouse to scroll the list and to click on *income93* as the Y-axis variable. Similarly, enter *baplus* as the X-axis variable. Click on Draw.
   **DOS:** In answer to the questions, the Y-axis variable should be `income93` and the X-axis variable, `baplus`.

The result of the procedure is shown in Figure 10.9.

   The regression line drawn on the scatter plot in Figure 10.9 seems to fit the overall pattern of the data fairly well. The equation for the line is printed at the top of the graph. It is *income93* = 8905.44 + 559.57*baplus*. This is a compact way of saying: "The expected value of a state's income per capita in 1993 conditional on that state's level of educational attainment is $8,905.44 plus 559.57 times the percentage of the state's residents 25 years or older who held at least a BA degree in 1990."

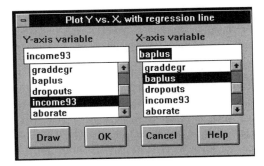

**Figure 10.8** *Dialog Box for* Plot Y vs. X, with Regression Line

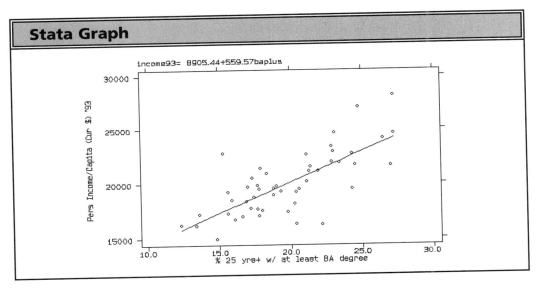

**Figure 10.9** *Output of* Plot Y vs. X, with Regression Line

The regression line produced by this equation gives the predicted value of *income93* for each state based on its level of *baplus*. The slope of the regression line is 559.57. This means a difference between states of one percentage point in *baplus* is associated with a difference of about $560 in income per capita. The slope is positive, so the higher the level of educational attainment, the higher the predicted income per capita.

Notice that there is some "scatter" of data points around the regression line. Evidently, a state's income per capita depends on other factors besides its level of educational attainment. Some states (like Nevada) are located well above the line; others (like Utah) are located well below it. These differences between predicted and actual values of income per capita are called residuals. They receive special attention in SQ's regression diagnostics (Chapter 17).

# Plot Y vs. X, Scale Symbols to Z

This scatter plot of Y vs. X uses circles as plotting symbols, but scales them to be proportional in relative size (area) to a third variable, which we'll call Z. The Z variable you choose can be any quantitative variable in your data set. (Don't try names or other string variables; you'll get an error message.) In the example below, Z is *pop1993*, a state's 1993 population in thousands. This type of graph provides yet another way to introduce a third variable into the analysis of Y vs. X.

**Example:** Based on our graphical analysis thus far, we've learned that *income93* and *baplus* are strongly and positively correlated in a linear relationship that can be summarized visually in a regression line, as shown in Figure 10.9. But there's still quite a lot of interstate variation in income per capita that can't be explained by educational differences. We hypothesize that a second "cause" variable might be population size. It could be that residents in the more populous states have higher earnings, on average, than those in the less populous states—and that this is true no matter what a state's level of educational attainment might be.

We test this idea visually by plotting *income93* versus *baplus* and scaling the plotting symbols proportional to each state's population size. If our reasoning is correct, we should see bigger circles at the top of the scatter (above the now-invisible regression line) and smaller circles at the bottom of the scatter (below the line).

**1**   From the *File* menu, choose *Open*.

**2**   Enter the name of the file: states3.

**3**   From the *Graphs* menu, choose *Scatterplots* and then *Plot Y vs. X, scale symbols to Z*.

**4**   **Windows:** In the dialog box (see Figure 10.10), scroll the list and click on *income93* as the Y-axis variable. Similarly, enter *baplus* as the X-axis variable and *pop1993* as the variable to scale points (Z). Click on Draw.

   **DOS:** In answer to the questions, the Y-axis variable should be income93, the X-axis variable, baplus, and the Z variable, pop1993.

The result of the procedure is shown in Figure 10.11.

In Figure 10.11, states with small populations are plotted with smaller circles, those with larger populations with larger circles. Can you spot Vermont? California? The picture is interesting but it does not offer any visual confirmation of our hypothesis. Little circles seem rather randomly mixed in with the big ones; there is no clear pattern, at least not the one predicted.

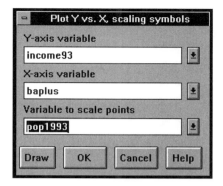

**Figure 10.10** *Dialog box for* Plot Y vs. X, scale symbols to Z

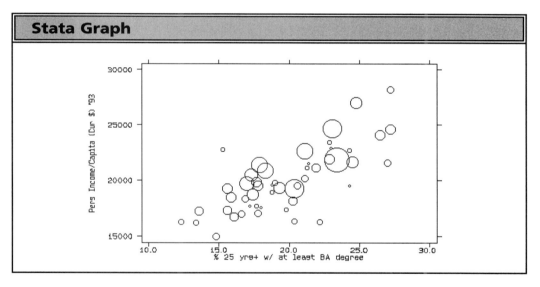

**Figure 10.11** *Output of* Plot Y vs. X, scale symbols to Z

## Plot Y vs. X, by Group

This procedure allows you to repeat a scatter plot analysis of X and Y for different subgroups of your data. It will show you the graphical results for all subgroups in one image. In the following example, the subgroups are four regions: Northeast, Midwest, South, and West. The output will be four separate scatter plots of *income93* vs. *baplus*, one plot for each region. All plots will have the same Y and X scales to make comparisons easier. If desired, a scatter plot of pooled data for all subgroups ("Total") will be shown along with the individual subgroup plots. This provides a graphical tool for analyzing a bivariate relationship while "controlling for" a third variable.

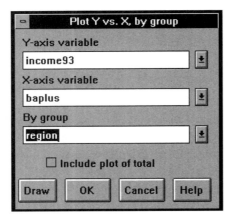

**Figure 10.12** *Dialog Box for* Plot Y vs. X, by Group

**Example:** We want to know whether the positive and linear relationship we found between income and education in the 50 U.S. states also holds true within each of the major regions of the country. If it does, we can be more confident that the relationship is a general one applicable to all states.

**1** From the *File* menu, choose *Open.*

**2** Enter the name of the file: states3.

**3** From the *Graphs* menu, choose *Scatterplots* and then *Plot Y vs. X, by groups.*

**4** **Windows:** In the dialog box (see Figure 10.12), scroll the list and click on *income93* as the Y-axis variable. Similarly, enter *baplus* as the X-axis variable, and *region* as the "By" variable. Do not click on the option box to include a plot of total. (Try this on the next round if you would like to see the result.) Click on Draw.

**DOS:** In answer to the questions, the Y-axis variable should be income93, the X-axis variable, baplus, and region should be the "By" variable. The plot of total menu option is not available in the DOS version.

The result of the procedure is shown in Figure 10.13.

Comparing the separate scatter plots for the four regions, we can see in Figure 10.13 that the relationship between income and education found earlier using data for all 50 states holds reasonably well for three of the four regions. In the plot for Western states, however, there is a real "scatter" of data points with no obvious pattern linking *income93* and *baplus* in this region. On closer inspection, it appears that a single state (the one all by itself in the upper left of the plot) is spoiling the linear pattern. Put your thumb over that state and the positive linear relationship between *income93* and *baplus* pops back into view for the states that remain. That "spoiler" state is

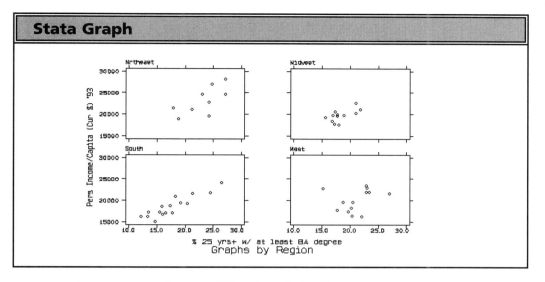

**Figure 10.13** *Output of* Plot Y vs. X, by Group

Nevada, which we spotted earlier as an odd duck relative to most other states. If we were to pursue this study, we might make a note to try robust regression (Chapter 18) or some other method designed to minimize the distorting effects caused by these kinds of "outliers" or "wild" values.

## Scatter Plot Matrix

SQ's *Scatter Plot Matrix* creates a panel of individual scatter plots of Y vs. X (and also X vs. Y) for all pairs of variables specified in a variable list. By viewing this collection of small scatter plots as an ensemble, you can see how the relationship varies between a particular variable of interest and each of the other variables included in the analysis. The individual plots shown in the scatter matrix have a lower resolution than those shown in larger scale by the standard Y vs. X plot—and the more variables plotted, the lower the resolution. Enough detail will usually be revealed, however, to show general patterns of relationship among the variables plotted. By scanning the matrix you can quickly spot bivariate relationships that seem particularly intriguing and worth studying in the larger and higher resolution standard scatter plot.

**Example:** As a preliminary step in a study of income per capita, high school dropout rates, and population density in the 50 U.S. states, we decide to look at a scatter matrix of relationships among these three variables.

**1**   From the *File* menu, choose *Open.*

**2**   Enter the name of the file: `states3`.

**3** From the *Graphs* menu, choose *Scatterplots* and then *Scatter plot matrix*.

**4** **Windows:** In the dialog box (see Figure 10.14), scroll the list and click on *income93*, *dropouts*, and *popdense* as the variables to be plotted. Click on Draw.

**DOS:** In answer to the questions, the variables to be plotted should be `income93`, `dropouts`, and `popdense`.

The result of the procedure is shown in Figure 10.15.

The easiest way to begin interpreting this matrix is to focus on the upper right triangle of scatter plots. Notice the variable names listed in the diagonal of squares that runs from upper left to lower right. Pick one of these as the Y variable of interest and pick another lower on the diagonal as the X variable of

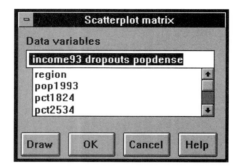

**Figure 10.14** *Dialog Box—Scatterplot Matrix*

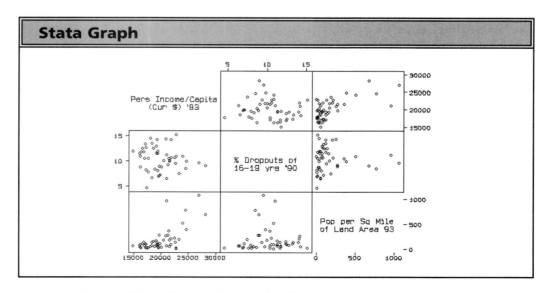

**Figure 10.15** *Output, Scatterplot Matrix*

interest. Imagine a cross-hair formed by a horizontal line across the Y-variable row and a vertical line up the X-variable column. The center of the cross-hair will fall on the Y vs. X plot you want to see. For example, if you pick *income93* as the Y variable and *popdense* as the X variable, the center of the cross-hair will fall on the plot of *income93* versus *popdense* in the upper right corner of the matrix. Notice that the mirror-image of this plot (*popdense* versus *income93*) appears in the lower left corner of the matrix. It shows exactly the same information, except that the original Y and X variables are reversed.

---

## ▓ TIP 12 ▓ ■ ■ ■ ■ ■ ■ ■ ■ ■ ■ ■ ■ ■ ■ ■

### Scatter Plot Command Mode Options

StataQuest provides a number of plotting symbol options for scatter plots. Here are some of them:

| | | | |
|---|---|---|---|
| O | (Upper case O) | Option syntax: | s(O) |
| o | (Lower case O) | Option syntax: | s(o) |
| . | (Period) | Option syntax: | s(.) |
| Δ | (Large triangle) | Option syntax: | s(T) |
| □ | (Large square) | Option syntax: | s(S) |
| ◊ | (Small diamond) | Option syntax: | s(d) |
| + | (Plus sign) | Option syntax: | s(p) |
| | (Invisible symbol) | Option syntax: | s(i) |
| abc | (String variable) | Option syntax: | s([string var name]) |

**Note:** The invisible symbol is impossible to display here for obvious reasons. It is used to instruct StataQuest to *suppress* display of plotting symbols that otherwise would be displayed. It is used, for example, in drawing clean straight regression lines on scatter plots.

**Note:** In StataQuest's graph command, symbol options are typed after the single comma that separates command and option fields. Here are three examples of such command statements. All produce a scatter plot of Y vs. X, each with a different symbol:

```
graph Y X, s(T)      [Symbol is a large triangle]
graph Y X, s(p)      [Symbol is a plus (+)]
graph Y X, s([city]) [Symbol is the city name stored in a variable
                      called city. Note: The square brackets are re-
                      quired around the string variable name.]
```

---

▪ **TIP 13** ▪ ▪ ▪ ▪ ▪ ▪ ▪ ▪ ▪ ▪ ▪ ▪ ▪ ▪ ▪ ▪

## Setting Text Size

Before creating a scatter matrix containing many small plots, switch to command mode and use the `set textsize 150` command to increase the size of the label text, thus making the output more readable and easier to interpret. The default textsize is 100. A textsize of about 150 (150%) will usually suffice for a matrix of six or seven variables. When you're finished, you'll probably want to restore text-size to its default value by entering the command `set textsize 100`.

---

▪ **TIP 14** ▪ ▪ ▪ ▪ ▪ ▪ ▪ ▪ ▪ ▪ ▪ ▪ ▪ ▪ ▪ ▪

## Scatter Matrices with a Period (.) as the Plotting Symbol

Sometimes a scatter matrix has so many plots packed with so many data points that no clear patterns can be discerned because the default plotting symbols overlap and cram the available space. In this situation, you can produce the desired scatter matrix in command mode using the period ( . ) as a plotting symbol. To illustrate, here is the command for creating the scatter matrix shown in Figure 10.15 with periods rather than lower case o's as plotting symbols:

```
graph income93 dropouts popdense, matrix s(.)
```

---

## ▪ EXERCISES

**10.1** Open the *lowbw.dta* file, which contains data collected for a sample of 189 mothers and their babies in a hospital study of low infant birth weights (Hosmer and Lemeshow, 1989). The Y variable of interest is *bwt*, infant birth weight in grams. The X variables to focus on are *age* (mother's age in years) and *lwt* (mother's weight in pounds at last menstrual period). You will also consider another variable, *smoke*, scored 1 if the mother smoked during pregnancy, 0 otherwise. Run the appropriate scatter plots to visually test the following hypotheses:

a. The heavier the mother, the heavier the baby.

b. The older the mother, the heavier the baby.

c. The older the mother, the heavier the baby for non-smokers—but not for smokers.

**10.2** Open the *birds.dta* file, which has various measurements on 32 species and subspecies of birds (McNab, 1994).

a. Plot *basalrat* (basal metabolism rate in cubic cm O2/h) versus *bodymass* (body mass in grams). What conclusions do you draw from this plot regarding the direction, strength, and functional form of the relationship between these two variables?

b. Run the same plot again, this time with a regression line. Would you say the linear model fits the data? Why or why not?

c. (More advanced) Run *Y vs. X plot, with regression line* again, only this time use *log10br* as the Y variable and *log10bm* as the X variable. Compare these results using the base-10 log transforms of the two variables with those obtained in (b).

**10.3** Open the *states3.dta* file, which contains various data for the 50 U.S. states.

a. Plot *aborate* (number of abortions per 1000 women 15–44 years old in 1992) versus *pctchris* (percentage of population identified as Christian adherents in 1990). What is the overall pattern of relationship between these two variables?

b. Run another scatter plot, this time using *abvstate* to name points. What does this add to your interpretation of the results in (a)?

c. Run a final scatter plot, this time scaling points proportional to *popdense* (population per square mile of land area in 1993). What pattern do you see in the results and how would you explain it?

**10.4** Open the *highway3.dta* file, which has various 1993 statistics on highway travel behavior and motor vehicle accidents in the 50 U.S. states. The argument is made that youth, speed, and crowding cause accidents and deaths on the highways. Produce a scatter plot matrix for the variables *fatalld* (number of motor vehicle accident fatalities per thousand licensed drivers), *pctun20* (percentage of licensed drivers under the age of 20), *exceed65* (percentage of drivers who exceed 65 miles per hour in 55 mph zones), and *popdense* (population per square mile of land area). What evidence do these graphical results lend to the argument?

# 11

# *Time Series Plots*

## Graphing Trends and Change

Time series plots are a special type of scatter plot in which a Y-axis measurement variable is plotted against an X-axis time variable (such as years, quarters, months) or an index (such as the trial number in an ordered sequence of experiments or the sample number in an ordered sequence of quality control checks). The adjacent data points plotted in a time series graph are usually connected by lines to make it easier to detect trends and fluctuations in levels of the Y variable over time. You can also graph multiple time series by plotting two or more Y variables against a single time or index variable.

SQ's *Time series* menu offers a choice of four different kinds of time series plots, as pictured in Figure 11.1:

- *Plot Y vs. X* will plot one Y variable against an interval-scaled X variable (that is, a time variable measured in constant intervals, such as days, weeks, years).

- *Plot more than one Y vs. X* will plot as many as six Y variables against an interval-scaled X variable.

| ■ ■ ■ ■ ■ ■ ■ | ■ | **Plot Y vs. X** |
| --- | --- | --- |
| **IN THIS CHAPTER** | ■ | **Plot more than one Y vs. X (without scaling option)** |
| | ■ | **Plot more than one Y vs. X (with scaling option)** |
| | ■ | **Plot Y vs. obs. no.** |
| | ■ | **Plot more than one Y vs. obs. no.** |

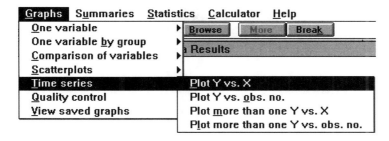

**Figure 11.1** *Menu Options for Graph-Time series*

■ For Y variables measured at irregular intervals over time, *Plot Y vs. obs. no.* will plot one Y variable against its order number in a sequence of such measurements.

■ *Plot more than one Y vs. obs. no* will plot as many as six Y variables.

For plots involving two or more Y variables measured on different scales, a graph option allows you to convert them to a common 0–100% scale defined by each variable's minimum and maximum value.

## Plot Y vs. X

This graph plots a single Y variable against an X variable that measures time in fixed intervals (for example, days, months, or years).

**Example:** We wish to study the trend in U.S. defense spending over the years 1962–1994.

**1** From the *File* menu, choose *Open.*

**2** Enter the name of the file: defense.

**3** From the *Graphs* menu, choose *Time series* and then *Plot Y vs. X*.

**4** **Windows:** In the dialog box (see Figure 11.2), scroll the list and click on *natldef* (annual defense budget outlays in $ billions) as the Y-axis variable. Similarly, enter *year* as the X-axis variable. Click on Draw.

**DOS:** In answer to the questions, the Y-axis variable should be `natldef` and the X-axis variable, `year`.

The result of the procedure is shown in Figure 11.3.

The time-series plot in Figure 11.3 shows that defense spending grew at a moderate rate over the 1960s and 1970s, accelerated over the 1980s, and then plateaued in the early 1990s.

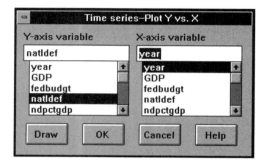

**Figure 11.2** *Dialog box for* Plot Y vs. X

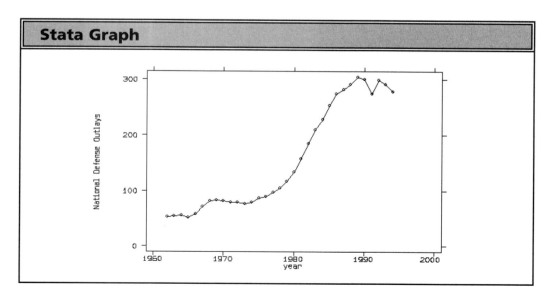

**Figure 11.3** *Output for* Plot Y vs. X

# Plot More Than One Y vs. X (without Scaling Option)

This graph plots two or more Y variables (as many as six) against an X variable that measures time in fixed intervals. The data points for each series will be marked by different plotting symbols and connected by distinct lines, allowing easy comparison.

**Example:** Continuing the defense spending example above, it would be interesting to compare the trend line shown for *natldef* in Figure 11.3 with one for the variable *ndpctgdp*, which measures defense spending as a percentage of annual gross domestic product (GDP). The GDP is an overall index of the nation's economic productivity.

**1**  From the *File* menu, choose *Open*.

**2**  Enter the name of the file: `defense`.

**3**  From the *Graphs* menu, choose *Time series* and then *Plot more than one Y vs. X*.

**4**  **Windows:** In the dialog box (see Figure 11.4), scroll the list and click on *natldef* and *ndpctgdp* as the Y-axis variables. Similarly, enter *year* as the X-axis variable. Click on Draw.

   **DOS:** In answer to the questions, the Y-axis variables should be `natldef` and `ndpctgdp`, and the X-axis variable, `year`.

The result of the procedure is shown in Figure 11.5.

For purposes of comparing trends, the multiple time-series plot of *natldef* and *ndpctgdp* displayed in Figure 11.5 is not very informative. That is because these two Y-axis variables are measured on different scales. The first Y variable, *natldef*, ranges from $50.6 billion to $303.6 billion in the data. The second Y variable, *ndpctgdp*, ranges only from 4.3% to 9.6%. Only a single Y-axis scale

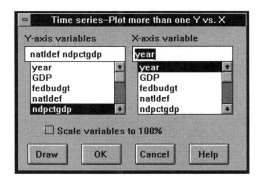

**Figure 11.4** *Dialog Box for* Plot more than one Y vs. X

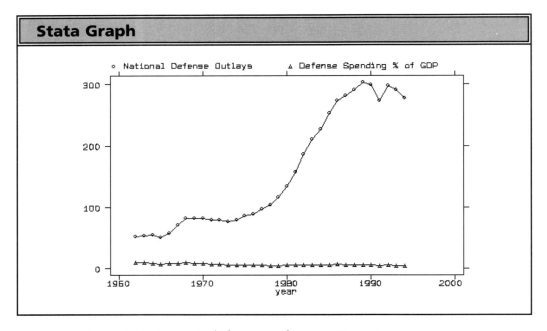

**Stata Graph**

**Figure 11.5** *Output of* Plot more than one Y vs. X

is available for plotting both variables, however, so it must include the lowest (4.3) and the highest (303.6) values of all variables in the graph. The visual result is that the time-series plot for *ndpctgdp* is compressed within a very small band of the total graph space. SQ's *Scale variables to 100%* option can fix this.

## Plot More Than One Y vs. X (with Scaling Option)

This plot will scale each Y variable to a range of 0 to 1.0, with 0 assigned to the variable's minimum value and 1.0 to the maximum value in the data. All Y variables plotted will then have the same 0 to 1.0 range.

**Example:** Again, we compare the trend line shown for *natldef* with the variable *ndpctgdp*.

**1** From the *File* menu, choose *Open*.
**2** Enter the name of the file: defense.
**3** From the *Graphs* menu, choose *Time series* and then *Plot more than one Y vs. X*.
**4** **Windows:** In the dialog box, scroll the list and click on *natldef* and *ndpctgdp* as the Y-axis variables. Similarly, enter *year* as the X-axis variable. This time click on the options box to scale variables to 100%. Click on Draw.

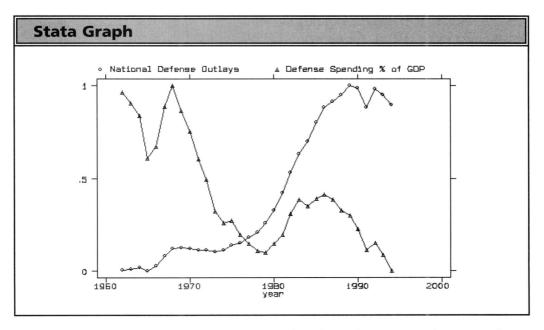

**Stata Graph**

**Figure 11.6** *Multiple Time-Series Plot of Trends in U.S. Defense Spending, 1962–1994.*

**DOS:** In answer to the questions, the Y-axis variables should be `natldef` and `ndpctgdp`; the X-axis variable, `year`. This time answer "yes" to the question about scaling the variables to 100%.

The result of the procedure is shown in Figure 11.6.

This rescaled version of Figure 11.5 makes it easier to see and compare trends in **both** time series. In general, spending for national defense increased over this period in actual budget outlays (circles), but decreased as a percentage of GDP (triangles).

## Plot Y vs. Obs. No.

This graph plots Y over time using the X axis to index the cumulative count of observations made in an ordered sequence of measurements. Examples are scores recorded for successive trials in an experiment (first trial, second trial, and so on) and measurements of successive samples of a product taken from an assembly line (first sample, second sample, and so on).

The difference between an X axis scaled as day one, day two, and so forth, and an X axis scaled as observation number one, observation number two, and so forth, is that in the latter case the time intervals between successive measurements of Y are not fixed and can vary. In an experiment, for example, the first trial might take place on day one, the second on day four, and the third on day nine. The *Plot Y vs. obs. no.* procedure allows you to chart trends in Y over a sequence of observations irregularly spaced in time.

**Example:** A student keeps meticulous records on the fuel consumption and efficiency of her 1990 Honda Civic. Her notebook entries of miles traveled, gallons consumed, and miles per gallon (mpg) for the 61 gas fill-ups between 5 November 1989 and 27 July 1991 are contained in the file *mpg.dta*. We want to study the trend in her car's fuel efficiency over this period.

The Y variable of interest is *mpg*. It is not necessary to specify an X variable. SQ assumes that X is the observation number *_n* (a temporary system variable used to keep track of the current order of cases in the file). This is not a problem because the data set is ordered ("sorted") on the variable *fillup*, which indexes the order of fillups from first (*fillup* = 1) to last (*fillup* = 61).

**1** From the *File* menu, choose *Open.*
**2** Enter the name of the file: mpg.
**3** From the *Graphs* menu, choose *Time series* and then *Plot Y vs. obs. no.*
**4** **Windows:** In the dialog box (see Figure 11.7), scroll the list and click on *mpg* as the Y-axis variable. Notice that there is no X-axis variable to select. Click on Draw.

   **DOS:** In answer to the question, the Y-axis variable should be mpg; there is no X-axis variable.

The result of the procedure is shown in Figure 11.8.

Figure 11.8 reveals that *mpg* fluctuated quite a bit from fillup to fillup over the period studied. This variability appears to have increased after about the 20th fillup. Further, we detect a slight overall downward trend in fuel efficiency from that point on. Seeing these results, a quality control expert might

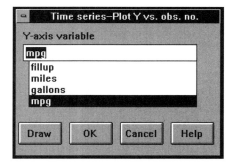

**Figure 11.7** *Dialog Box for* Plot Y vs. obs. no.

**Stata Graph**

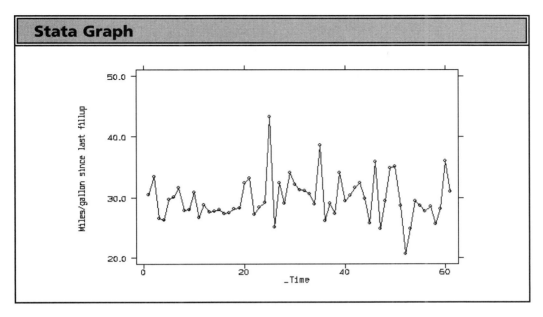

**Figure 11.8** *Output of* Plot Y vs. obs. no.

conclude that the fuel efficiency maintenance process was starting to go "out of control."

**Caution:** Before running these types of time-series plots, be sure the **order** of observations on Y is correct. Sometimes the original order of your data can be changed by using other procedures. As a result, what you thought was the first observation of Y might now be the sixth or tenth. You can check this by listing your data (*List data* from the *Summaries* menu) to make sure the observation numbers are consistent with the Y values shown. If they're not, use the *Sort* command in SQ's *Editor* to straighten things out.

## Plot More Than One Y vs. Obs. No.

This graph plots two or more Y-axis variables (as many as six) against an observation number. The graph option for scaling variables to 100% is also available for this procedure.

**Example:** The *gears1.dta* file contains data for 30 successive samples of four gears, each taken from a production process. The observation number therefore ranges from 1 (first sample) to 30 (last sample). The outside diameter of each gear was measured and recorded, yielding four measurements for each sample. The variable *lowmsmt* stores the smallest diameter recorded in each sample, and the variable *highmsmt* stores the largest diameter. These are the two Y-axis variables to be plotted against the observation (sample) number.

---

▨ **TIP 15** ▨ ▨ ▨ ▨ ▨ ▨ ▨ ▨ ▨ ▨ ▨ ▨ ▨ ▨

### Restoring the Original Order of Your Data

If it is important to be able to restore the original order of your data, and if you don't have a variable (for example, sample number, experiment number) in your data set storing the original observation numbers, then before you do anything with the data set, you should create and save such a variable. One way to do this is to enter `generate obsno = _n` in SQ's command window and then save and replace the file. If at a later time your data have been reshuffled into a different order, you can then easily restore the original order by sorting on *obsno* in the spreadsheet or by entering `sort obsno` in command mode.

---

**1** From the *File* menu, choose *Open.*

**2** Enter the name of the file: `gears1`.

**3** From the *Graphs* menu, choose *Time series* and then *Plot more than one Y vs. obs. no.*

**4** **Windows:** In the dialog box (see Figure 11.9), scroll the list and click on *lowmsmt* and *highmsmt* as the Y-axis variables. Notice again that there is no X-axis variable to select. Click on Draw.

**DOS:** In answer to the questions, the Y-axis variables should be `lowmsmt` and `highmsmt`; there is no X-axis variable.

The result of the procedure is shown in Figure 11.10.

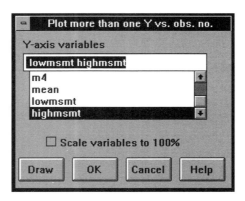

**Figure 11.9** *Dialog Box for* Plot more than one Y vs. obs. no.

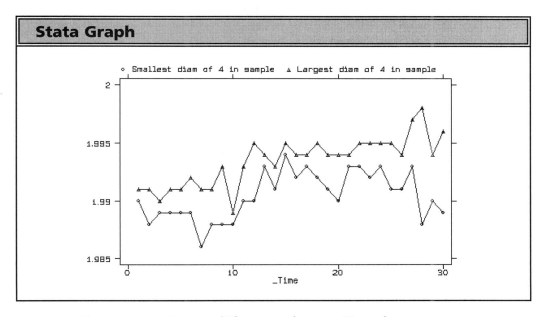

**Figure 11.10** *Output of* Plot more than one Y vs. obs. no.

In Figure 11.10, circles are used to plot *lowmsmt* and triangles to plot *highmsmt* over the 30 samples. The gap between these measurements for any given sample is equal to the range (highest value minus lowest value) for that observation. You can see that this range, a measure of variability, appears to increase dramatically in the last samples taken. This trend would alarm a quality control analyst (see Chapter 12), who would probably take steps to discover the cause.

## ▓ EXERCISES

**11.1** The *airbombs.dta* file contains time-series data on security screening, detection of weapons, and arrests at U.S. airports over the period 1978–1991.

   a. Plot the number of passengers screened each year over the period studied.

   b. In one graph, plot the number of handguns, longguns, bombs, and other weapons detected each year over this period. Do you see any pattern in the type of weapon most often detected? Any trends?

   c. In one graph, plot the number of passengers screened each year and the arrest rate per million screenings for carrying bombs and firearms. (Did you use the "scale variables to 100%" option?) Compare these two time series and suggest how they might be linked.

**11.2** The *gearsall.dta* file contains the data for samples 1–30 used in the "gears" example in the chapter text. It also contains the data for 30 new samples (31–60) taken after modifications were made to the production process (Smith, 1947). Using the appropriate time-series graph, plot *lowmsmt* and *highmsmt* over the entire sequence of 60 samples. What evidence is there in this plot that the modifications achieved a better result?

**11.3** We have all heard about the baby boom that began in the United States immediately following World War II. One measure used in studying such trends is the general fertility rate, which is the ratio of live births to the total number of women aged 18 to 44. In the *fertilty.dta* file, the variable *fertrate* records the general fertility rate each year over the period 1930–1990 (Wright, 1989, 1992). Plot *fertrate* versus *year*. In what year does the fertility rate reach its peak? Based on the graphical evidence, how would you describe the trend in fertility rates before and after World War II?

**11.4** The *crime.dta* file contains various U.S. annual crime statistics for the period 1974–1993.

   a.   Plot the annual crime index rate (*indrate*) versus *year*. Describe the pattern or trend observed. Using the variable *pctviol*, is there evidence that crimes became more violent over this period?

   b.   In one graph, plot the annual violent crime index rate and the property crime index rate over this period. Is rescaling needed to compare these two time series? How would you describe these two series and what comparative conclusions would you draw?

# *Quality Control Charts*

## What Are Quality Control Charts?

Managers have used quality control charts since the 1920s to maintain high standards in manufacturing processes. In all production processes, there is variation from ideal standards in what comes out at the end of the assembly line. Such variation can be large or small, random or nonrandom, controllable or uncontrollable, acceptable or unacceptable. Acceptable variation usually means small, random, and uncontrollable. To prevent unacceptable variation, quality control specialists keep production processes "in control" by sampling and inspecting output, monitoring rejection rates, taking measurements, computing statistics, and preparing quality control charts.

There are several types of control charts. All display summary statistics for sample measurements with clearly marked lower and upper limits defining the range of acceptable variation. StataQuest's **C chart** displays information on product defects. The **P chart** focuses on rejection rates. The **X-bar chart** and **R chart** show sample means and ranges, respectively, for measurements of product characteristics. The **X-bar + R Chart** combines these last two charts in one image.

---

▨ ▨ ▨ ▨ ▨ ▨

## IN THIS CHAPTER

■ **Control (C) charts for defects**

■ **Fraction defective (P) charts**

■ **X-bar charts**

■ **Range (R) charts**

■ **X-bar + R charts**

---

## Control (C) Chart for Defects

The C chart is used to control the number of defects ("nonconformities") found in standard inspection units of production. Examples of defects are accounting errors found in samples of 100 financial records, flaws counted in 3 square feet of wood surface in sampled table tops, or errors counted in samples of 1,000 lines of computer code. Note in each example that the inspection unit does not vary: 100 records, 3 square feet of surface, 1,000 lines of code. Each count has the same sample weight.

Typically, the analyst samples N inspection units, records the number of defects in each, and calculates the mean number of defects (C), the upper control limit (UCL), and the lower control limit (LCL). A Y versus X control chart is then constructed, with the Y axis measuring the number of defects and the X axis indexing inspection units from the first to the Nth unit sampled. Horizontal lines are drawn from the Y axis at the estimated values for UCL and LCL. The counts of defects are then plotted over the sequence of samples 1 through N. The plotting symbols are connected with a line to help detect trends. Sampled units that fall above the UCL line or below the LCL line are identified as out of control.

**Note:** The theoretical sampling distributions underlying these and other charts are discussed in standard texts on quality-control methods. One such text is Farnum (1994), which is also the source of the *cchart.dta* and *pchart.dta* data files used later in this chapter.

**Example:** The *cchart.dta* file contains counts of defects observed on each of 50 circuit boards randomly sampled from a production process. We construct a C chart to look for trends and to identify any units that are out of control.

**1**   From the *File* menu, choose *Open*.

**2**   Enter the name of the file: `cchart`.

**3**   From the *Graphs* menu, choose *Quality control* and then *Control (C) chart for defects*.

**4** **Windows:** Click on the variable named *defects* as the defects count variable and on the variable named *board* as the identification variable.

**DOS:** Enter `defects` as the defects count variable and `board` as the identification variable.

The results appear in Figure 12.1.

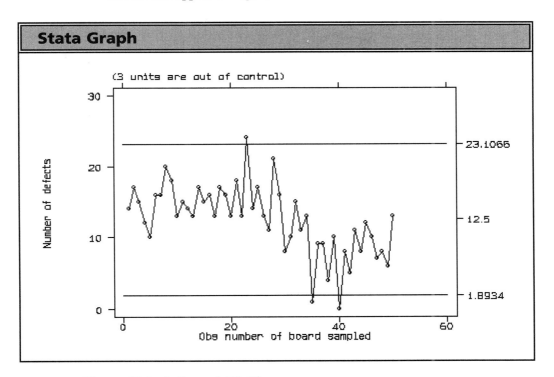

**Figure 12.1** *A Control (C) Chart*

Figure 12.1 plots the number of defects found in each circuit board over the sequence of 50 samples. Shown are horizontal lines giving the estimated LCL (1.8934) and UCL (23.1066). On the right scale we see that the mean count of defects for the 50 inspected boards is 12.5. Three units are identified as out of control—two below the LCL and one above the UCL. This C chart also reveals a downward trend (that is, improvement) in the number of defects starting around sample 30, with a swing back toward the mean during the last inspections.

## Fraction Defective (P) Charts

The P chart is used to control the **fraction** of defective items in a production process. Samples (subgroups) of items are inspected over fixed time intervals (for example, daily). For each sample the number of items tested and the number of rejects are recorded. If the number of items tested in each sample varies, the computed LCL and UCL also will vary and the plotted LCL and UCL lines will have a ragged appearance. (The *stabilize* option in command mode will standardize the data and draw straight lines. See Tip 17 in this chapter.) The P chart resembles the C chart and is used in the same way to detect out-of-control samples and overall trends.

**Example:** The *pchart.dta* file contains data for 30 samples of circuit boards, one such sample taken daily over a 30-day period. Each sample is identified by the day number (*day*) in sequence (1, 2, ... 30). The number tested (*tested*) and the number defective (*rejects*) are recorded for each day's sample. The number of boards tested ranges from 281 to 328 over the 30 samples. Because the sample size varies, we expect to see ragged UCL and LCL lines.

**1**   From the *File* menu, choose *Open*.

**2**   Enter the name of the file: `pchart`.

**3**   From the *Graphs* menu, choose *Quality control* and then *Fraction defective (P) chart*.

**4**   **Windows:** Click on the variable named *rejects* as the variable for number of defectives, on the variable named *day* as the identification variable, and on the variable named *tested* as the sample-size variable.

   **DOS:** Enter `rejects` as the variable for number of defectives, `day` as the identification variable, and `tested` as the sample-size variable.

The results appear in Figure 12.2.

The P chart in Figure 12.2 shows that the process for producing circuit boards is under control for the period studied. No units are out of control. Fluctuation occurs in the fraction defective observed over time, but the variation appears random and well within the lower and upper control limits.

## X-Bar Charts

X-bar charts are used to achieve statistical control over the mean of a measured characteristic of a product. For example, a manager might desire statistical control over the mean outside diameter of a gear measured to the nearest ten-thousandth of an inch. Another manager might want control

## Stata Graph

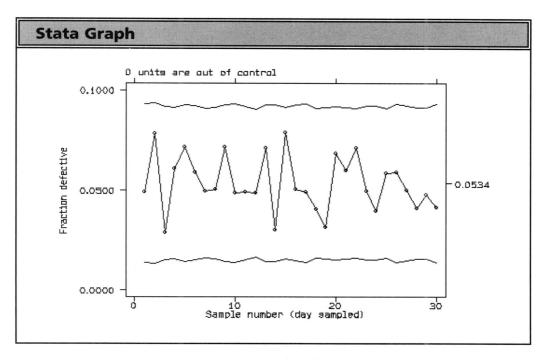

**Figure 12.2** *A Fraction Defective (P) Chart*

over the mean concentration of electroless nickel in plating solutions in ounces per gallon.

X-bar charts and R charts (see following page) are usually used together in studying a process. The R (range) chart is typically used to establish reliable estimates of confidence intervals (UCL and LCL) around the sample means (X-bars). To identify the desired level of control, a horizontal "control line" is drawn in the X-bar chart at the grand mean (the mean of sample means) estimated from the data. Other specified norm values can also be used to set the control line.

The data used in X-bar charts is collected for N subgroups of K items sampled during a production period. For X-bar charts to be useful, at least 25 subgroups should be sampled and K should be held constant at 4, 5, or 6 items per subgroup.

Given sets of measurements on the K items within each of the N subgroups, the quality control analyst can use SQ to compute the subgroup means and ranges, the grand mean of the subgroup means, and the mean of the subgroup ranges.

---

### ▦ TIP 16 ▦ ▦ ▦ ▦ ▦ ▦ ▦ ▦ ▦ ▦ ▦ ▦ ▦ ▦ ▦

### Stabilizing UCL and LCL Lines in P Charts

The wavy UCL and LCL lines in P charts result from varying sample sizes. They can be "stabilized" as constants by transforming the fraction defective to standard deviation units (Z-scores). Stabilization is achieved by using StataQuest's *pchart* command with the *stabilize* option. To produce a stabilized version of the P chart shown in Figure 12.2, for example, type `pchart rejects day tested, ylab xlab stabilize` in SQ's Command window and press the ENTER key. This command and option will draw the UCL and LCL as straight-line constants at +3.0 and -3.0 standard deviation units, respectively.

---

**Example:** During World War II, a manufacturing firm produced a small gear used in a weapon. Quality control analysts randomly sampled 30 subgroups of four gears each from the production process. They measured each gear's outside diameter to the nearest ten-thousandths of an inch. In the *gears1.dta* file, the variable *sample* indexes the subgroup (1, 2, 3, ... 30), and the variables *m1*, *m2*, *m3*, and *m4* record measurements of the first, second, third, and fourth gear, respectively, in each subgroup. We'll use this information to create an X-bar chart.

1   From the *File* menu, choose *Open*.

2   Enter the name of the file: `gears1`.

3   Press the **Editor** button, place the cursor on the *sample* column, and click the **Sort** button.

   **Alternative:** Use Command mode or the Command window: type `sort sample` in the Command window and press the ENTER key.

4   From the *Graphs* menu, choose *Quality control* and then *X-bar chart*.

5   **Windows:** Click on each of the variables *m1*, *m2*, *m3*, and *m4* to enter them in the variable list box as the sample measurement variables.

   **DOS:** Enter `m1`, `m2`, `m3`, and `m4` as the sample measurement variables.

The results appear in Figure 12.3.

The X-bar chart in Figure 12.3 tells an interesting story (Smith, 1947, 57–66). The outside diameters of gears in the first subgroups sampled were much too small relative to the control line norm of 1.99185 inches shown on the right-axis scale. Three subgroups were below the LCL of 1.98949 inches

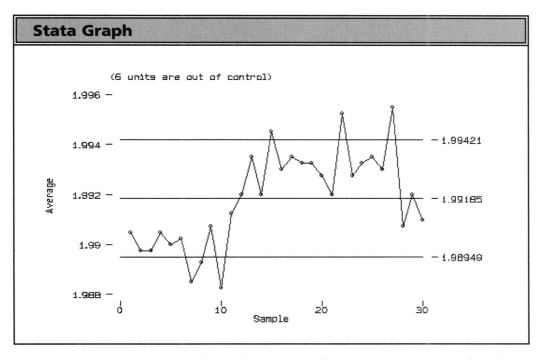

**Figure 12.3** *X-Bar Chart of Mean Outside Diameters in 30 Samples of Four Gears Each*

and thus out of control. The cause was traced to a machine operator who deliberately ran the process low to reduce material that had to be removed from the grinding. After correcting that problem and resuming sampling, the process went out of control in the other direction, with three subgroups exceeding the UCL of 1.99421 inches. The cause of the problem in this instance was traced to a worn bearing in one of the machines. After repair, the production process seemed to stabilize around the control line. At that point, data were collected for a second set of 30 subgroups and new charts were drawn. (See Exercise 12.1.)

## Range (R) Charts

R charts are similar in design to X-bar charts and are based on the same measurement data. Instead of plotting subgroup **means** over the sequence of samples, however, they plot subgroup **ranges** (largest value minus smallest value). R charts are used to study variability in the production process, to detect trends, and to establish confidence intervals for the LCL and UCL lines used in X-bar charts. The R chart has its own control line and LCL and UCL lines. If the population standard deviation is unknown or unspecified, SQ will estimate control limits for both the X-bar chart and the R chart by using the mean of subgroup ranges in combination with formula weighting factors that are published in standard textbooks.

**Example:** We continue our study of the gears data, this time asking for an R chart.

**1**   From the *File* menu, choose *Open*.

**2**   Enter the name of the file: `gears1`.

**3**   Press the **Editor** button, place the cursor on the *sample* column, and click the Sort button.

  **Alternative:** Use Command mode or the Command window: type `sort sample` in the Command window and press the ENTER key.

**4**   From the *Graphs* menu, choose *Quality control* and then *Range (R) chart*.

**5**   **Windows:** Click on each of the variables *m1*, *m2*, *m3*, and *m4* to enter them in the variable list box as the sample measurement variables.

  **DOS:** Enter m1, m2, m3, and m4 as the sample measurement variables.

The results appear in Figure 12.4.

Figure 12.4 shows a trend toward increasing within-sample variability in the last subgroups sampled. Two of the last three subgroup ranges are identified as out of control. This trend developed just as the subgroup means were being brought under statistical control (see Figure 12.3).

## X-Bar + R Charts

X-bar charts and R charts are best studied in tandem; the information in one illuminates the other. SQ's X-bar + R chart plot makes it easy to look at them together. To illustrate, we'll reproduce the output of the last two graph commands in one simple procedure.

**1**   From the *File* menu, choose *Open*.

**2**   Enter the name of the file: `gears1`.

**3**   Press the Editor button, place the cursor on the *sample* column, and press the Sort button.

  **Alternative:** Use Command mode or the Command window: type `sort sample` in the Command window and click the ENTER key.

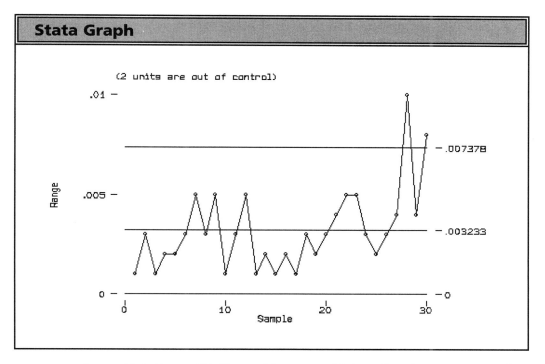

**Figure 12.4** *Range (R) Chart of Ranges in 30 Samples of Four Gears Each*

**4**   From the *Graphs* menu, choose *Quality control* and then *X-bar & R chart*.

**5**   **Windows:** Click on each of the variables *m1*, *m2*, *m3*, and *m4* to enter them in the variable list box as the sample measurement variables.

   **DOS:** Enter m1, m2, m3, and m4 as the sample measurement variables.

The results appear in Figure 12.5.

   As shown in Figure 12.5, the X-bar + R chart makes it easier to compare the two charts displayed earlier in Figures 12.3 and 12.4. The two charts in Figure 12.5 plot different sample statistics (means and ranges) against the same sample index number. If a particular subgroup were found to be out of control on the X-bar chart, for example, an eyeball check of the R chart below it might reveal an extreme value for the range in that subgroup, which in turn might be the result of a single outlier gear. X-bar + R charts provide an over-all snapshot of the data, whereas the individual X-bar and R charts offer tools for more detailed study.

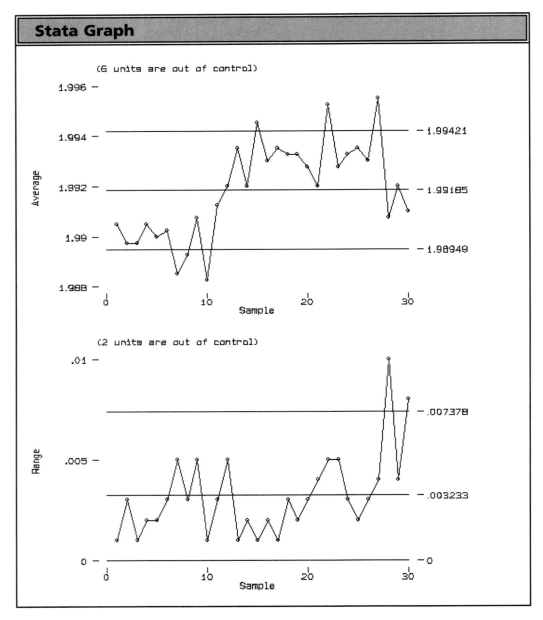

**Figure 12.5** *X-Bar + R Chart Plot Summarizing Output of Figures 12.3 and 12.4*

## ■ TIP 17 ■ ■ ■ ■ ■ ■ ■ ■ ■ ■ ■ ■ ■ ■ ■ ■

### Additional Options for R Charts and X-Bar Charts

In constructing an R chart, if you know the standard deviation of the population ("process") distribution from which units are sampled, you can specify that value directly by typing the *rchart* command with the *std(#)* option in SQ's Command window. In an analysis of the gears data, for example, if the population standard deviation were known to be .002, you could type `rchart m1 m2 m3 m4, std(.002)` in the Command window and press the ENTER key. SQ would then use that input to calculate LCL and UCL lines rather than estimate them from the sample data.

The same option is available for SQ's *xchart* command. Three additional options for the *xchart* command are *mean(#)*, *lower(#)*, and *upper(#)*. The *mean(#)* option allows you to specify a known or desired grand mean. Use the *lower(#)* option to specify the LCL **and** the *upper(#)* option to specify the UCL (both must be specified, not just one). Continuing the gears example, to construct an X-bar chart with the grand mean set at 1.993 and the LCL and UCL set at 1.991 and 1.995, respectively, type `xchart m1 m2 m3 m4, mean(1.993) lower(1.991) upper(1.995)` in the Command window and press the ENTER key.

**Note:** As with all SQ commands in command mode, the individual options can be specified in any order. Of those discussed here, only the *mean(#)* and *std(#)* options are available for X-bar + R charts.

## ■ EXERCISE

**12.1** To continue the gear manufacturing firm case, the control chart analysis discussed in the text led to modifications of the production process. The firm's researchers then randomly sampled another 30 subgroups of four gears each, took the same kinds of measurements, and did another analysis. These data are contained in the *gears2.dta* file. Question: Did the modifications succeed in bringing the production process under statistical control? Your answer should be supported, minimally, by the evidence of an X-bar chart and an R chart along with your interpretation of the results.

# 13

# Parametric Tests

## What are Parametric Tests?

**Parametric** tests are tests "whose derivations involve explicit assumptions about population distributions and parameters" (Hays and Winkler, 1971, 778). These are often assumptions that data or parameters are distributed according to a normal distribution; another assumption used is that variances are known or equal. Data for parametric tests are often required to be interval or ratio level—such as age, income, height, or weight—as opposed to such categorical or ordinal variables as religion or public opinion scales from strongly agree to strongly disagree. Tests that don't involve assumptions about the population distribution and can be used with categorical and ordinal level data are called **nonparametric** and are considered in the next chapter.

In this chapter we cover the commonly used t-tests, one-sample and two-sample tests of proportion, as well as the less commonly used equality of variance test and two tests of normality.

## One-Sample t-Test

A single sample t-test tests the hypothesis that the population mean of some variable is equal to some number. StataQuest will ask you what number the

■ ■ ■ ■ ■ ■ ■
## IN THIS CHAPTER

- ■ **One-sample t-test**
- ■ **Two-sample t-test**
- ■ **Paired t-test**
- ■ **One-sample and two-sample tests of variance**
- ■ **Testing for normality**
- ■ **One-sample test of proportion**
- ■ **Two-sample test of proportions**

mean should be tested against. The test assumes that the population from which the sample mean is drawn is normally distributed and uses the t distribution because the true population variance and standard deviation are unknown and are being estimated with the sample standard deviation. Kanji (1993, 27) states that the test will be an "approximate guide" if the underlying population is not normally distributed.

The data set *bearings.dta* contains a random sample of 20 ball bearings randomly picked from two production lines at the end of the day (Hand, 1994, 131). Both production lines are set to produce ball bearings with a diameter of 1 micron. The variable *diameter* contains the actual diameters of the 20 ball bearings. Our null hypothesis is that the diameter of the ball bearings produced equals 1 micron. The alternative hypothesis is that the diameters do not equal 1 micron. Our procedure will be to load the data into memory, test the sample of 20 against the null hypothesis, and then divide the data into two subsets of 10 and test each. We will use both procedures outlined in Chapter 3, *Modifying Your Data*, to divide the data into subsets. In the next section, we will test the hypothesis that the two production lines are producing equal sized ball bearings.

**1**   From the *File* menu, choose *Open*.

**2**   Enter the name of the file: `bearings`.

**3**   From the *Statistics* menu, choose *Parametric tests* and then *1-sample t test*.

**4**   **Windows:** For the variable, click on *diameter*. The hypothesized mean should be 1. Click on OK.

 **DOS:** In answer to the questions, the variable should be `diameter`, and the hypothesized mean should be 1. The confidence level is 95.

The result is shown in Figure 13.1.

In Figure 13.1, StataQuest first tells us the Stata command it executes: `ttest diameter = 1`. Then it tells us the variable (diameter), the number

```
  Stata Results

. ttest diameter = 1

  Variable |       Obs         Mean    Std. Dev.
  ---------+--------------------------------
  diameter |        20          1.3     .3721205

            Ho:   mean = 1
                    t = 3.61 with 19 d.f.
             Pr > |t|  = 0.0019
```

**Figure 13.1** *One-Sample t-Test*

of observations (20), the mean of those 20 observations (1.3), and their standard deviation (.3721205). It then restates the null hypothesis, tells us the t-value (3.61), degrees of freedom (19), and then how likely the observed sample result or one more extreme is if the null hypothesis is true (0.0019). Thus our t-value of 3.61 means that the probability that a sample with a mean diameter of 1.3 or more could have arisen by chance from a population where the average ball bearing diameter is 1 micron is less than 0.01. Conclusion: something may be wrong with one or both production processes.

Perhaps one production line is malfunctioning, or perhaps both are. To distinguish between the two production lines, we will run the same one-sample t-test on each production line. To do this, we must take two subsets of the data. In Chapter 3, *Modifying Your Data*, we discussed two ways of doing this: one way is to retrieve the previous command in command mode (Method 1; Windows only) and add an "if" statement to it; the other way is to save two subsets of the data (Method 2).

## Method 1: Retrieving the Previous Command (Windows Only)

■ We remember from Chapter 3 that we need to write an "if" statement. The statement should refer to the production line variable, which we don't remember. We decide to run the command *Describe variables*: From the *Summaries* menu, choose *Describe variables*. The variable is *prodline*.

■ We then decide to make sure we know how *prodline* is coded. From the *Summaries* menu, we choose *Tables, 1-way (frequency)*, and then indicate in the dialog box that the variable is *prodline*. The results are in Figure 13.2, indicating that *prodline* is coded with a 1 for the first production line and a 2 for the second.

■ We then know our "if" statement: if prodline == 1.

■ We then use the PGUP key three times to recall the previous commands. If we go too far, we can use PGDN to move down the list. When we have `ttest diameter = 1` in the Command window, we add our "if" statement so that the command reads `ttest diameter = 1 if prodline == 1`. We then press ENTER to execute the command.

The result is shown in Figure 13.2.

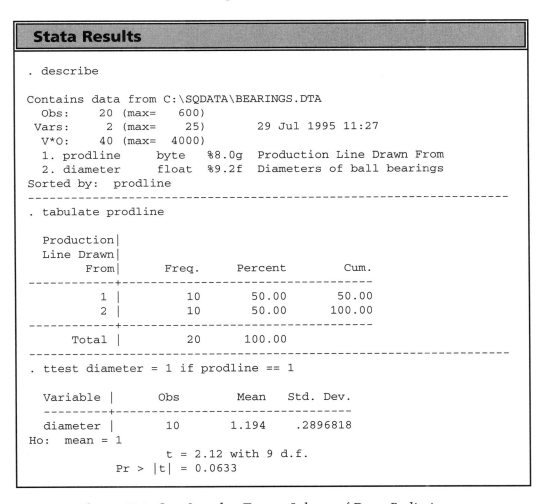

```
Stata Results

. describe

Contains data from C:\SQDATA\BEARINGS.DTA
  Obs:     20  (max=   600)
  Vars:     2  (max=    25)       29 Jul 1995 11:27
  V*O:     40  (max=  4000)
  1. prodline      byte    %8.0g   Production Line Drawn From
  2. diameter      float   %9.2f   Diameters of ball bearings
Sorted by:  prodline
-----------------------------------------------------------------
. tabulate prodline

  Production|
  Line Drawn|
        From|     Freq.       Percent        Cum.
  -----------+----------------------------------------
          1 |       10         50.00         50.00
          2 |       10         50.00        100.00
  -----------+----------------------------------------
      Total |       20        100.00
-----------------------------------------------------------------
. ttest diameter = 1 if prodline == 1

  Variable |     Obs         Mean     Std. Dev.
  ---------+------------------------------------
  diameter |      10        1.194     .2896818
Ho:   mean = 1
                 t = 2.12 with 9 d.f.
           Pr > |t| = 0.0633
```

**Figure 13.2** *One-Sample t-Test on Subsets of Data, Preliminary Information and Subset 1*

The results for `ttest diameter = 1 if prodline == 1` indicate that we fail to reject the null hypothesis that the mean = 1.

We then use the PGUP key to recall the `ttest diameter = 1 if prodline == 1` command and change the 1 at the end to 2. The results are in Figure 13.3. It appears there that the culprit is production line #2, where the average diameter of the ball bearings is 1.406, and that the probability of our drawing a sample with that mean or greater from a population where the average diameter equals 1 is less than 0.015. We conclude that production line #2 needs adjustment.

---

### Stata Results

```
. ttest diameter = 1 if prodline == 2

 Variable |      Obs       Mean    Std. Dev.
----------+-------------------------------
 diameter |       10      1.406    .4283093

         Ho:  mean = 1
                 t = 3.00 with 9 d.f.
          Pr > |t| = 0.0150
```

---

**Figure 13.3** *One-Sample t-Test on Subsets of Data, Subset 2*

## Method 2: Save Subsets of Data

**1** From the *File* menu, choose *Open*.

**2** Enter the data set name: `bearings`.

**3** Enter the Editor by clicking on the **Editor** button (Windows) or selecting *Spreadsheet* from the *Edit* menu (DOS).

**4** Place the cursor on the *prodline* variable in any of the observations where *prodline* is equal to 1.

**5** **Windows:** Click on the **Delete** button. From the dialog box, choose *Delete all 10 obs. where prodline == 1*. Click on OK.

**6** **Windows:** Exit the Editor by clicking on the **Close** button. Click on OK to accept changes (that were made in the Editor). From the *File* menu, choose *Save As* and give the new dataset a new name, such as `bear3`.

**DOS:** Press F10 to activate the spreadsheet menu; from the *Drop* menu, choose *Drop observations*. SQ will tell you it is about to delete all 10 observations where prodline is equal to 1. Press ENTER to indicate that this is appropriate. Choose *Save* from the *File* menu and give the data set a new name: `bear3`.

**7**   Follow the same procedure to drop the 10 observations where prodline ==
2; save this data set as `bear4`.

**8**   From the *File* menu, choose *Open*. Enter the name *bear3* to read the 10
observations into memory where prodline is equal to 1. Do the one-
sample t-test on this data set. Then do the same with *bear4*.

Your results should look like Figure 13.4.

---

## Stata Results

```
. use C:\SQDATA\BEARINGS.DTA, clear
. edit
- preserve
- drop if prodline == 1

. save C:\SQDATA\BEAR3.DTA
file C:\SQDATA\BEAR3.DTA saved

. use C:\SQDATA\BEARINGS.DTA, clear

. edit
- preserve
- drop if prodline == 2

. save C:\SQDATA\BEAR4.DTA
file C:\SQDATA\BEAR4.DTA saved

. use C:\SQDATA\BEAR3.DTA, clear

. ttest diameter = 1
        ...Results omitted-same as Figure 13.2

. use C:\SQDATA\BEAR4.DTA, clear

. ttest diameter = 1
        ...Results omitted-same as Figure 13.2
```

**Figure 13.4** *Method 2: One-Sample t-Test on Subsets of Data*

In Figure 13.4, note that when you enter the Editor, StataQuest writes
. edit on the Results window; when you exit and allow the changes made in
the Editor to become part of the data set in memory, SQ writes - preserve
in the Results window. When you dropped the observations where the variable
*prodline* was equal to 2, SQ used its own command language to document that
action, writing in the Results window: - drop if prodline == 2.

## Two-Sample t-Test

The two-sample t-test differs from the one-sample t-test in that the two-sample test assumes that we are testing the mean of one variable against the mean of another variable from the same data set. (The nonparametric equivalent of the two-sample t-test is the Mann-Whitney test; see Chapter 14.)

StataQuest first makes us select an option. Which of these is chosen depends on the structure of the data. We will use the previous data set to illustrate. The two options presented are listed above the Tables 13.1 and 13.2.

**Table 13.1** *One Data Variable, One Group Variable*

| prodline | diameter |
|----------|----------|
| 1 | 1.18 |
| 1 | 1.42 |
| 1 | 0.69 |
| 1 | 0.88 |
| 2 | 1.72 |
| 2 | 1.62 |
| 2 | 1.69 |
| 2 | 0.79 |

**Table 13.2** *Two Independent Data Variables*

| diampr1 | diampr2 |
|---------|---------|
| 1.18 | 1.72 |
| 1.42 | 1.62 |
| 0.69 | 1.69 |
| 0.88 | 0.79 |

In Table 13.1, we have listed the first four observations from production line #1 and the first four observations from production line #2. When StataQuest asks you whether the option is "1 data variable, 1 group variable," SQ assumes you have the data arranged so that there is a group variable like production line (prodline) and then a second variable with the data, arranged such that the data from group 1 is next to a 1 in the group variable column, and so on.

On the other hand, when SQ asks you whether the data are arranged for "2 independent data variables," SQ is assuming that instead of having a group variable, you have two separate columns in your data matrix, as in Table 13.2. Notice that the specific data points are the same in both tables. Note especially that even though *diampr1* and *diampr2* are both on the same row or observation, *diampr1* and *diampr2* are not related to each other. We could take either variable and mix up the data points; there is no significance attached to the fact, for example, that 1.18 and 1.72 are on the same row; 1.18 and 1.62 could be in the first row of data, and we would achieve the same results. The only difference between Tables 13.1 and 13.2 is how we as data analysts have arranged the data. [We are making this point strongly because in the Paired Difference test, there **is** explicitly such a relationship-see the next section.]

We can now carry the analysis of the ball-bearing production lines one step further, asking whether there is a significant difference between production line #1 and production line #2. In the last section, we found that the mean and standard deviation for production line #1 was 1.19 and 0.29 respectively; for production line #2, 1.41 and 0.43. So there is a difference: 1.19 is not the same number as 1.41. Our question is whether this difference is statistically significant; were the two samples drawn from two distinct populations (each population being a production line)? Or could this difference have arisen just because these are two different random samples drawn from the same population? The two-sample t-test will provide the evidence.

Note that in the *bearings.dta* file, our data is arranged as in Table 13.1.

**1** From the *File* menu, choose *Open*.

**2** Enter the name of the file: bearings.

**3** From the *Statistics* menu, choose *Parametric tests* and then *2-sample t test*.

**4** **Windows:** In the first dialog box, click on *1 data variable, 1 group variable*; then click OK. In the second dialog box, click on *diameter* as the data variable; click on *prodline* as the binary (2 categories) group variable. Then click OK.

**DOS:** In answer to the questions, the data are arranged as Option 1. The data variable should be diameter, and the group variable should be prodline. Assume equal variances and a 95% confidence level.

The result is shown in Figure 13.5.

```
Stata Results

. use C:\SQDATA\BEARINGS.DTA, clear

. ttest diameter, by(prodline)

Variable |     Obs       Mean    Std. Dev.
---------+---------------------------------
       1 |      10      1.194    .2896818
       2 |      10      1.406    .4283093
---------+---------------------------------
combined |      20        1.3    .3721205

Ho:  mean(x) - mean(y) = 0 (assuming equal variances)
                   t = -1.30 with 18 d.f.
           Pr > |t| = 0.2112
             95% CI = (-.55552801, - 13152799)
```

**Figure 13.5** *Two-Sample t-Test*

The results show that we fail to reject the null hypothesis that the two production lines are not significantly different from each other. Notice that we ignored the question about unequal variances. Because the standard deviations of the two production lines seem quite different (0.29 vs. 0.43), we rerun the test choosing the unequal variance option. The results (not shown) are essentially the same (the t-values are the same; the degrees of freedom are different; the significance level is the same to the hundredths place).

## Paired t-Test

The paired t-test is used to test the possibility that two variables in the same data set are related to each other; here we assume that the variables are matched and come from the same observation. We will test the possibility, a remote one if the ordinary person's view of campaigns and campaign finance is correct, that the winner's campaign expenditures are equal to the loser's, using the real 1994 campaign election data from a random sample of the 1995 U.S. House of Representatives, *hrsamp94.dta*. (The nonparametric equivalents of the paired t-test are the Sign test and the Wilcoxon signed-ranks test, covered in Chapter 14.)

**1** From the *File* menu, choose *Open*.

**2** Enter the name of the file: `hrsamp94`.

**3** From the *Statistics* menu, choose *Parametric tests* and then *Paired t test*. (DOS: Choose *two-sample t-test*.)

**4** **Windows:** In the dialog box, variable #1 should be *expwin*; variable #2 should be *explose*. Then click OK.

**DOS:** In answer to the questions, our data are paired (option 3), variable 1 should be `expwin`; variable 2 should be `explose`. Confidence level is again 95.

The result is shown in Figure 13.6.

The results indicate virtually no chance (less than 0.00005) that we could have obtained these results by chance by sampling from a population distribution where the two means were equal. We conclude that there is a substantial ("statistically significant") difference between the average amount that winners spend and the amounts that their opponents spend.

## One-Sample and Two-Sample Tests of Variance

The one-sample test of variance tests the standard deviation of a variable in your data set against a hypothesized standard deviation that you specify, similar to the one-sample t-test. The two-sample test of variance test assumes that your data

```
Stata Results

. ttest expwin = explose

  Variable |      Obs         Mean    Std. Dev.
 ---------+----------------------------------
    expwin |      108     585.0991     370.5739
   explose |      108     295.9685     518.7016
 ---------+----------------------------------
     diff. |      108     289.1306     559.6572

          Ho:  mean difference = 0   (paired data)
                             t = 5.37 with 107 d.f.
                    Pr > |t| = 0.0000
          95% CI for difference = (182.37356,395.88756)
```

**Figure 13.6** *Paired t-Test*

are divided into two subsets. You want to test the possibility that the standard deviation (the square root of the variance) of a variable in one subset is the same as the standard deviation of that variable in the other subset. It is **not** necessary that the means of the two groups be equal. The test assumes that the populations from which the samples are drawn follow normal distributions.*

We will test the possibility that the variances in the distributions from the two ball bearing production lines are equal. Recalling Figure 13.5, production line #1 had a standard deviation of 0.29; production line #2, 0.43.

**1** From the *File* menu, choose *Open*.

**2** Enter the name of the file: bearings.

**3** From the *Statistics* menu, choose *Parametric tests* and then *2-sample text of variance*.

**4** **Windows:** In the dialog box, the data variable is *diameter*; the binary group variable is *prodline*. Then click OK.

   **DOS:** In answer to the questions, the data variable is diameter; the binary group variable is prodline.

The result is shown in Figure 13.7.

Our null hypothesis is that the two standard deviations are equal. Our results show that there is a high probability (0.2595) that our results arose by chance. We fail to reject the null hypothesis.

---

* See Kanji (1993) for a discussion of the assumptions underlying the 100 most common statistical tests.

```
┌──────────────────────────────────────────────────────────────┐
│ Stata Results                                                  │
├──────────────────────────────────────────────────────────────┤
│ . sdtest diameter, by(prodline)                               │
│                                                                │
│   Variable |      Obs        Mean    Std. Dev.                 │
│   ---------+------------------------------------               │
│          1 |       10       1.194    .2896818                  │
│          2 |       10       1.406    .4283093                  │
│   ---------+------------------------------------               │
│   combined |       20         1.3    .3656258                  │
│                                                                │
│          Ho:  sd(1) = sd(2)   (two-sided test)                 │
│   Lower tail:  F1(9,9) =  0.46                                 │
│   Upper tail:  F2(9,9) =  2.19                                 │
│     (Pr < F1) + (Pr > F2) =  0.2595                            │
│                                                                │
└──────────────────────────────────────────────────────────────┘
```

**Figure 13.7** *Equality of Variance Test*

# Testing for Normality

In addition to the individual coefficients for skewness and kurtosis, StataQuest contains two tests of normality, the Shapiro-Wilk W and Shapiro-Francia W'. W is appropriate for sample sizes between 7 and 2,000 observations, and W' is appropriate for sample sizes from 5 to 5,000. The null hypothesis is that the values of a variable have been drawn from a normal distribution. The alternative hypothesis is that the values of the variable have not been drawn from a normal distribution. These tests should be used in conjunction with the skewness and kurtosis coefficients in Chapter 4, *Summarizing and Examining Your Data*. There is a discussion of the various tests of normality in issues 1–5 of the *Stata Technical Bulletin* and Judge et al. (1985).

**1**  From the *File* menu, choose *Open*.

**2**  Enter the name of the file: `bearings`.

**3**  From the *Statistics* menu, choose *Parametric tests* and then *Normality test*.

**4**  **Windows:** In the dialog box, the variable is *diameter*. Then click OK.

   **DOS:** In answer to the question, the variable is `diameter`.

The result is shown in Figure 13.8.

In Figure 13.8, the Shapiro-Wilk W test prints out under the command "swilk diameter," and the Shapiro-Francia W' test prints out under the command "sfrancia diameter." The probability of these results having arisen by chance is high at 0.715 and 0.890, and we fail to reject the null hypothesis that the distribution of ball bearing diameters is normal.

```
Stata Results

. swilk   diameter

                   Shapiro-Wilk W test for normal data
    Variable |    Obs          W          V          z     Pr > z
    ---------+------------------------------------------------------
    diameter |     20      0.96814      0.754     -0.569   0.71519

. sfrancia   diameter

                   Shapiro-Francia W' test for normal data
    Variable |    Obs          W'         V'         z     Pr>z
    ---------+------------------------------------------------------
    diameter |     20      0.98090      0.500     -1.230   0.89058
```

**Figure 13.8** *Shapiro-Wilk W and Shapiro-Francia W' Tests of Normality—Ball Bearing Diameters*

On the other hand, some data are quite different. Consider, for example, the distribution of populations among the 130 nations of the world in the *humandev.dta* file. The variable in question is *pop1988*. The results are in Figure 13.9. Here we reject the null hypothesis that the distribution is normal.

```
Stata Results

. swilk   pop1988

                   Shapiro-Wilk W test for normal data
    Variable |    Obs          W          V          z     Pr > z
    ---------+------------------------------------------------------
    pop1988  |    130      0.27543     74.616     9.703    0.00000

. sfrancia   pop1988

                   Shapiro-Francia W' test for normal data
    Variable |    Obs          W'         V'         z     Pr>z
    ---------+------------------------------------------------------
    pop1988  |    130      0.26421     82.523     8.243    0.00001
```

**Figure 13.9** *Shapiro-Wilk W and Shapiro-Francia W' Tests of Normality—Populations of the Nations of the World*

# One-Sample Test of Proportion

A single sample test of proportion uses a sample proportion to test a hypothesis about a population proportion. The only requirement is that the sample be a random one, that is, that the items in the sample be chosen independently from one another. Each should have the same probability of falling into one of the categories of interest.

The data set *gsssurv.dta* contains a random sample of the American public, including a question about whether the respondent voted in 1992. (For more information about the survey, use the *Dataset information* command in the *Summaries* menu.) The survey indicates that 70.73% of the sample voted in 1992, representing 406 of the 574 persons in the survey. We can test the hypothesis, using the one-sample test of proportion, that more than two-thirds of the population voted. Are our sample results significantly higher than two-thirds of the public? Or could our sample results have been drawn from a population where two-thirds of the public state that they have voted?

## Binomial Test

SQ gives you the choice of running a binomial test or using the normal approximation for the binomial test. For the binomial test, SQ needs several items of information:

■ "Data variable [0/1]," which is the variable in question. The variable must be coded in two categories, 0 and 1. Our data variable is *vote92*.

■ "Probability of success or exp. no. successes," which is the population probability in the null hypothesis against which we are going to test our sample data. This item can be expressed as a probability (between 0 and 1) or in terms of the number of successes. Our "probability of success" is .66666, or we can calculate the number of successes, which would be .66666 times the sample size, which is 574. 574 x 0.66666 = 382.66.

■ A "Conf. level" or Confidence level, for which the default is 95, which means 95%. Confidence levels must be more than 0 and less than 1.

■ "Normal approximation" is a box that can be checked to substitute the normal approximation for the binomial distribution.

To do a binomial test, follow these steps:

**1** From the *File* menu, choose *Open*.

**2** Enter the name of the file: `gsssurv`.

**3** From the *Statistics* menu, choose *Parametric tests* and then *1-sample test of proportion*.

**4**  **Windows:** For the Data variable, click on *vote92*. The probability of success should be .66666. Leave the "Conf. level" alone; do not check "Normal approximation." Click on OK.

**DOS:** In answer to the questions, the variable to be tested is `vote92`. The probability of success is `.66666`. Leave the "Sig. level" on the default of 95.

The result is shown in Figure 13.10.

## Stata Results

```
. bintest vote92=.66666, level(95)

    Varname |   Obs. N    Obs. k    Exp. k    Assumed p    Obs. p
   ---------+----------------------------------------------------------
     vote92 |     574       406    382.6628    0.66666     0.70732

       [1]  Pr(k>=406)                = 0.020741   (one-sided test)
       [2]  Pr(k<=406)                = 0.983383
       [3]  2*min([1],[2])            = 0.041483   (two-sided test)

       [4]  Pr(k==406)                = 0.004125
       [5]  Pr(k==360)                = 0.004767
       [6]  Pr(k==359)                = 0.003991
       [7]  Pr(k>=406 | k<=360)       = 0.046361   (two-sided test)

Ho: proportion(x) = .66666

                                             -- Binomial Exact --
   Variable |     Obs       Mean    Std. Err.  [95% Conf. Interval]
   ---------+----------------------------------------------------------
     vote92 |     574   .7073171    .0189911   .6682903    .7442597
```

**Figure 13.10** *One-Sample Test of Proportion*

This output in Figure 13.10 is the most complex in StataQuest. We have 574 observations ("Obs. N"), with 406 stating that they have voted ("Obs. K"). We expected to find 382.6628 on the basis of the hypothesized two-thirds ("Exp. k" and "Assumed p"). And the 406 represents 0.70732 of the 574-person sample size ("Obs. p"). StataQuest then gives us our output:

**Line 1.** A one-tailed test of the probability of observing 406 or **more** successes. This is an unlikely probability (0.02).

**Line 2.** The probability of getting 406 or **fewer** successes. If our observed successes had been less than the expected successes (382.66), then Line 2 would have been labeled as the "(one-tailed test)."

**Line 3.** A crude two-tailed test that is twice the one-tailed test on Line 1 or 2. Here it is twice Line 1. A more sophisticated two-tailed test is derived in Lines 4 through 6 and contained on Line 7.

**Line 4.** The probability of exactly k (here 406) number of successes.

**Lines 5 and 6.** Part of the process of deriving the two-tailed test probability. Here we have 406 successes on one side of the distribution. What would be the corresponding levels on the other side of the distribution? StataQuest looks at both 360 successes and 359 successes and prints them in Lines 5 and 6.

**Line 7.** The actual two-tailed test probability. Only if the assumed probability of success in the population is 0.50 (which it is **not** in our example) will Lines 3 and 7 be the same.

We conclude that our sample could not have been drawn from a universe where two-thirds of the public voted.

## Normal Approximation to the Binomial Test (Windows Only)

The information supplied is the same as running the binomial test, except that the box next to the words "Normal approximation" is checked. "Normal approximation" means that the test is similar to the binomial test, but is not the same as the binomial test. To use the normal approximation, the number of observations should be more than 30, or the expected number of observations in the smaller of the two categories should be 9 or more (Hamilton, 1992, 290–291). Figure 13.11 shows the results, which are simpler to interpret than the binomial test results above.

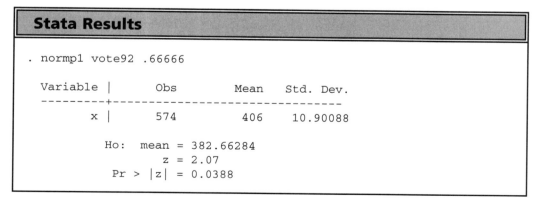

**Figure 13.11** *One-Sample Test of Proportions (Normal Approximation)*

The results are similar. We again reject the null hypothesis and conclude that our sample is unlikely to have come from a population where two-thirds of the public voted.

## Two-Sample Test of Proportions

The two-sample test of proportions can be used to compare two distributions when those distributions are composed of variables that take the form of a binomial distribution, that is, they are either a "success" or a "failure." Success and failure must be coded 1 and 0. The test compares the proportion of successes in the first sample with the proportion of successes in the second sample. The null hypothesis is that the difference between the two proportions is zero (for example, Ho: proportion(x) - proportion(y) = 0).

Like the two-sample t-test, SQ asks you first how your data are arranged, and again the choices are the following:

■ **One data variable, one group variable** (in the following example, we have a variable for the proportion of the public expressing a high degree of confidence in American business; we have a group variable for gender, male and female). See the example in Table 13.1.

■ **Two independent data variables.** See the example in Table 13.2.

The 1994 General Social Survey has a number of questions about the level of each respondent's confidence in various American institutions. Is the proportion of men expressing "a great deal of confidence" (as opposed to "no" confidence, or "some" confidence) the same as the proportion of women? We test that possibility:

**1** From the *File* menu, choose *Open*.

**2** Enter the name of the file: gsssurv2.

**3** From the *Statistics* menu, choose *Parametric tests* and then *2-sample test of proportions*.

**4** **Windows:** SQ first asks how the data are arranged; choose "One data variable, one group variable." Click on OK. A second dialog box then asks for "Data variable [0/1]," which should be *conbus*. The box also asks for "Group var. [2 groups]," which should be *sex*. Click again on OK.

  **DOS:** In answer to the questions, the data are arranged as "One data variable, one group variable" (option 1). The variable to be tested (0/1) is conbus; the group variable is sex. The confidence level is 95.

The result is shown in Figure 13.12.

```
  Stata Results

. prtest conbus, by(sex) level(95)

  Variable |       Obs   Proportion
  ---------+---------------------
         1 |       160      .29375
         2 |       235    .1957447
  ---------+---------------------
  combined |       395     .235443

     Ho:  proportion(x) - proportion(y) = 0
                                    z = 2.25
                            Pr > |z| = 0.0242
                  95% conf. interval = (0.0111,0.1849)
```

**Figure 13.12** *Two-Sample Test of Proportions*

The results in Figure 13.12 show that there appears to be a statistically significant difference between men and women on the probability of expressing a high degree of confidence in American business as an institution. About 29% of men (code 1) express a high degree of confidence (.29375); only about 20% of women (.195...) share this view.

Notice that SQ automatically runs the normal approximation for the binomial distribution for the two-sample test of proportions. The cautionary notes about sample size expressed earlier in the section on the one-sample test of proportions apply here also.

## ■ EXERCISES

**13.1**   Use the two-sample t-test to ascertain whether there is a significant difference between British Kings and Queens in their lifespans after taking office as compared with U.S. Presidents and Roman Catholic Popes. The data set is *pres.dta*; the lifespan variable is *years*, and the variable differentiating the groups is *kings*.

**13.2**   Use the one-sample t-test to ascertain whether the salaries of those who enter occupations in the marketing area could have been drawn from a sample with an average salary of $30,000. Do the same with those from the education area with the data set *salaries.dta*.

**13.3** Use the two-sample t-test to ascertain whether there is a significant differ-ence in birth weights between those who lived and those who died for infants with severe idiopathic distress syndrome. The data set is *respdist.dta*.

**13.4** Use the data from 13.3 to check the birth weight variable for normality and for equal variances in the two groups.

**13.5** The data set *birds.dta* has several items of information on 32 New Zealand birds. Use the two-sample t-test to ascertain whether the birds that fly are different from those that are flightless on their body mass, basal rate, and pectoralis.

**13.6** Use the data from problem 13.5 to check the body mass and basal rate vari-ables for normality and equal variances between the two groups.

**13.7** In the data set *gsssurv2.dta* are variables for the level of confidence the American public expresses in the following institutions: education, Congress, the scientific community, and television. Is the level the same for both men and women? Use the two-sample test of proportions provide evidence. How do your results compare with those in Figure 13.12? Why?

**13.8** Use the two-sample t-test to ascertain whether the salaries of those who enter occupations in the marketing area are different from those who enter the education area. The data set *salaries.dta* contains the entrance salaries advertised in days specializing in different occupations in the Manchester Guardian, and the salaries are listed in both British pounds and U.S. dollars.

**13.9** The data set *caffeine.dta* contains the results of an experiment where three groups of 10 male college students each had a certain dose of caffeine and then did a simple physical task, in this case, finger-tapping. Use the two-sample t-test to ascertain whether the evidence supports differentiating those who had the highest dose of caffeine (200 mg) from everyone else. The variable *group2* is coded 1 for those with the highest dose and 0 for everyone else.

**13.10** (Involves using subsets) Test whether the three groups of college students in *caffeine.dta* (problem 13.8) have equal variances in the number of finger taps per minute.

**13.11** The data set *crying.dta* contains the result of an experiment on the effects of rocking babies on their incidence of crying in a hospital. Use the two-sample t-test to ascertain whether there is a difference in the percentage of the babies crying each day. The relevant variables are *ctrlsucc* and *rocksucc*.

# 14

# *Nonparametric Tests*

## What Are Nonparametric Tests?

In this chapter, we cover the five nonparametric tests included in StataQuest: the Sign test, the Wilcoxon signed-ranks test, the Mann-Whitney test, the Kruskal-Wallis test, and the Kolmogorov-Smirnov test.

Where the data for parametric tests are often required to be interval or ratio level, the data for nonparametric tests can often be categorical or ordinal (in ranks). Where parametric tests frequently assume normal distributions, nonparametric tests make few assumptions about the population distribution. Nonparametric tests are sometimes called distribution-free in the sense that they do not require explicit assumptions of normality or homogeneity of variances. Instead, the assumptions are relatively "mild" ones.

## Sign Test

The Sign test is the oldest of the nonparametric tests, as well as one that uses very little information from the data. Consider the parametric t-test, which considers each data point and its distance from the mean. The Wilcoxon signed-ranks test, considered in the next section, substitutes ranks for the

▪ ▪ ▪ ▪ ▪ ▪

## IN THIS CHAPTER

- ■ **Sign test**
- ■ **Wilcoxon signed-ranks test**
- ■ **Mann-Whitney test**
- ■ **Kruskal-Wallis equality of populations test**
- ■ **Kolmogorov-Smirnov tests**
- ■ **Summary (chart)**

actual distances. The Sign test, considered in this section, goes even further and uses only the sign of the difference between the two variables. So it considers only whether X is greater than, equal to, or smaller than Y. The Sign test assumes only that the data are a random sample, each observation is independent, and that the data scale is at least ordinal and continuous. It is most comparable with the paired t-test among the parametric tests.

The data are assumed to be matched pairs, that is, the two variables on each row of the data matrix must refer to the same observation. If the two variables are A and B, then the null hypothesis is that the median of the difference between A and B is zero, that is, Ho: median(A-B) = 0. The test statistic is the number of observations for which A-B is greater than 0. The test statistic has a binomial distribution where n is the sample size and p is equal to 0.5.

Here are some data on the effectiveness of a new fuel additive in an experiment with 12 cars. The data are provided in *autoexp1.dta*. The variable for each car's performance without the additive is *perf*; the variable with the additive is *perfadd*. Figure 14.1 contains a list of the data with a variable, *diff*, defined as *perfadd* minus *perf*. In other words, *diff* is the change in mileage from using the additive.

Notice that most vehicles had better performance with the additive. Of the 12 observations, 8 increased, 1 remained the same, and only 3 went down. Our question is whether the 8 is significantly different from what would have been expected if the data were random and half went up and half went down. In other words, is the 8 significantly different from what we would get with p = 0.5, or 6 out of 12?

To run the Sign test:

**1**  From the *File* menu, choose *Open*.

**2**  Enter the name of the file: `autoexp1`.

**3**  From the *Statistics* menu, choose *Nonparametric tests* and then *Sign test*.

```
 Stata Results

. list
            perf     perfadd        diff
   1.        20          24            4
   2.        23          25            2
   3.        21          21            0
   4.        25          22           -3
   5.        18          23            5
   6.        17          18            1
   7.        18          17           -1
   8.        24          28            4
   9.        20          24            4
  10.        24          27            3
  11.        23          21           -2
  12.        19          23            4
```

**Figure 14.1** *Listing of the Automobile Additive Experiment Data*

**4** **Windows:** The Sign test dialog box asks for "Data Variable #1" and "Data Variable #2." Click on *perf* for the first and *perfadd* for the second. Click on OK.

**DOS:** In answer to the questions, the first variable to test should be `perf` and the second, `perfadd`.

The result is shown in Figure 14.2.

In Figure 14.2, as always, SQ first tells us the command it executed (`signtest perf=perfadd`). Next, it provides us with a table of the results when it compared *perf* with *perfadd*. In the table, 3 observed cases were positive (*perf* was greater than *perfadd*); 8 were negative (*perfadd* was greater than *perf*), and one was the same or zero. SQ excludes observations that have no difference between the two variables when it considers what it should "expect"; in this case, it expected 5.5 (or half of 11) to be in each category.

The table then gives us three significance tests. The first and second are one-sided; the third is two-sided. The alternative hypotheses are as follows: In the first, the alternative hypothesis is that *perf* is greater than *perfadd*; our experimental result is that 3 observations fell into this category. SQ then gives us the probability that we could have an experimental result of 3 when we are expecting 5.5, the sample size is 11, and p (from the Binomial distribution) is 0.5. The p-value is 0.9673. So the probability of getting a result of 3 or fewer by chance when we are expecting 5.5 in a Binomial distribution is very high at almost 0.97.

```
Stata Results

. signtest perf = perfadd

Sign test

    sign |    observed    expected
---------+-----------------------
positive |         3         5.5
negative |         8         5.5
    zero |         1           1
---------+-----------------------
     all |        12          12

One-sided tests:
Ho: median of perf = perfadd vs. Ha: median of perf > perfadd
    Pr(#positive >= 3)
    = Binomial(n = 11, x >= 3, p = 0.5) =  0.9673

Ho: median of perf = perfadd vs. Ha: median of perf < perfadd
    Pr(#negative >= 8)
    = Binomial(n = 11, x >= 8, p = 0.5) =  0.1133

Two-sided test:
Ho: median of perf = perfadd vs. Ha: median of perf ~= perfadd
    Pr(#positive >= 8 or #negative >= 8)
    = min(1, 2*Binomial(n = 11, x >= 8, p = 0.5)) =  0.2266
```

**Figure 14.2** *Sign Test Results Window*

SQ then includes the other one-sided test, that of *perf* being less than *perfadd*. Our experimental result is that we had 8 observations where *perfadd* was greater than *perf*, and our Binomial distribution has p again equal to 0.5, n = 11, and the probability that this result could arise by chance is 0.1133, a much smaller probability than 0.9673, but not small enough that we would normally reject the null hypothesis.

SQ also includes a two-sided test, which is not relevant for us in this particular experimental context. It has a similar interpretation. The probability that the result could have arisen by chance is 0.2266. So, although 8 of the 12 vehicles had a higher mileage, the difference was not substantial enough to be statistically significant.

# Wilcoxon Signed-Ranks Test

The Wilcoxon signed-ranks test tests the equality of distributions for matched pairs of observations. We have two related samples whose measurement scales allow us to consider whether X is different from Y in any single observation and also by how much X is different from Y. We determine the magnitude of the difference, then we rank the magnitudes. The Wilcoxon signed-ranks test uses those ranks to determine whether X is different from Y.

We assume that we can form a difference score, D, consisting of the difference between X and Y. Each pair of measurements, X and Y, is on a single observation or on subjects who have been paired on the basis of other variables. The sample should be random. The differences are assumed to be observations on a continuous random variable, symmetric, independent, and interval-level data (Daniel, 1978, 135). All observations where X is equal to Y (and thus there is a 0 difference score) are eliminated from the analysis. When the difference score is equal for more than one observation, each is given an average rank.

There is a small-sample version of the Wilcoxon signed-ranks test, but StataQuest runs the large-sample approximation, which is valid for samples larger than about 25 (Daniel, 1978, 138). The large-sample approximation forms a z-value that is distributed as a standard normal distribution and has the traditional interpretation.

In general, because the Wilcoxon signed-ranks test uses more information from the data than the Sign test, it is considered more powerful. However, it requires at least interval-level data, whereas the Sign test can operate with ranks or ordered categories.

The fuel-additive experiment used in the last section for the Sign test has only 12 observations and thus cannot be used for this test (because we need a "large" sample for the large-sample normal approximation). Instead, we shall compare terminal cancer patients who took vitamin C in large doses with control patients who did not. The data set is *paulch14.dta*, composed of 31 stomach and colon terminal cancer patients and the corresponding control group. Variables *timet1* (patient) and *timec1* (average of 10 matched controls) are the number of days the patient or the control group lived from the date of the first hospital attendance for the cancer that reached the terminal stage. Variables *timet2* (patient) and *timec2* (average of 10 matched controls) are the number of days the patient or control group lived from the dates of untreatability. If vitamin C in large doses makes a difference, then the treated patients should, on the average, live more days than the controls.

To run the Wilcoxon signed-ranks test:

**1** From the *File* menu, choose *Open*.

**2** Enter the name of the file: `paulch14`.

**3** From the *Statistics* menu, choose *Nonparametric tests* and then *Wilcoxon signed-ranks*.

**4** **Windows:** The Wilcoxon test dialog box asks for "Data Variable #1" and "Data Variable #2." Click on *timet1* for the first and *timec1* for the second. Click on OK.

**DOS:** In answer to the questions, the first variable to test should be `timet1` and the second, `timec1`.

**5** Do the same test for the variables from the time of the last hospital entry, *timet2* and *timec2* (not shown).

The result is shown in Figure 14.3.

```
Stata Results

. signrank timet1 = timec1

Wilcoxon signed-rank test

    sign |      obs   sum ranks    expected
---------+---------------------------------
positive |       23        437       218.5
negative |        0          0       218.5
    zero |        7         28          28
---------+---------------------------------
     all |       30        465         465

unadjusted variance      2363.75
adjustment for ties        -0.25
adjustment for zeros      -35.00
                       ----------
adjusted variance        2328.50

Ho: median of timet1 = timec1
           z =    4.528
    Prob > |z| =    0.0000
```

**Figure 14.3** *Wilcoxon Signed-Ranks Test*

The results in Figure 14.3 show a significant difference between the treatment group and the controls, that is, the probability of obtaining a z-value of 4.528 or greater is less than 0.00005.

# Mann-Whitney Test

The Mann-Whitney test, also known as the Mann-Whitney Two-Sample Statistic and the Wilcoxon Rank Sum test, is used to test the hypothesis that two samples are from populations with the same median or the same distribution. The test requires that the data be divided into two groups; the groups do not have to be the same size, nor is it assumed that they match in any way. The assumptions are that the two samples are random and independent, the data variable is continuous and random, the measurement scale is at least ordinal, and that the distributions differ only with respect to location (Daniel, 1978, 82).

The test statistic combines the two samples and ranks each data point from smallest to largest. Tied observations are given the average rank for the group. The ranks for the first population are then summed. This figure, S, is then adjusted by sample size and becomes U (Conover, 1980, 217). StataQuest runs a normal approximation to the Mann-Whitney test whereby the null hypothesis is that the two medians are equal, that is Ho: median(X1) – median(X2) = 0. The alternate hypothesis is that they are not equal.

We shall demonstrate the Mann-Whitney test on some data on the effectiveness of behavioral methods of psychotherapy. The data set is *prsscore.dta*, in which there are 17 subjects who improved their scores on Klopfer's Prognostic Rating Scale (PRS) for subjects who recently received behavior modification therapy, and 10 subjects who did not improve. Our question is whether these two distributions are the same.

To run the test:

**1** From the *File* menu, choose *Open*.

**2** Enter the name of the file: `prsscore`.

**3** From the *Statistics* menu, choose *Nonparametric tests* and then *Mann-Whitney*.

**4** **Windows:** The Mann-Whitney dialog box asks for "Data Variable " and "Group Var. (2 groups)." Click on *prsscore* for the data variable and *improve* for the group variable. Click on OK.

**DOS:** In answer to the questions, the variable to be tested should be `prsscore` and the group variable should be `improve`.

The result is shown in Figure 14.4.

In Figure 14.4, as usual, StataQuest first tells us its own command language: `. ranksum prsscore, by(improve)`. It then produces a table with a row for each category of *improve*, indicating the number of ranks, the sum of those ranks, and the sum expected if the two distributions were identical. SQ then has a row where it combines the samples, adding up each column. It then has an

```
Stata Results

. ranksum prsscore, by(improve)

Two-sample Wilcoxon rank-sum (Mann-Whitney) test

 improve |       obs    rank sum     expected
---------+------------------------------------
       1 |        17       296.5          238
       2 |        10        81.5          140
---------+------------------------------------
combined |        27         378          378

unadjusted variance       396.67
adjustment for ties        -0.24
                       ----------
adjusted variance         396.42

Ho: median prsscore(improve==1) = median prsscore(improve==2)
          z =    2.938
    Prob > |z| =    0.0033
```

**Figure 14.4** *Mann-Whitney U Test Demonstrated on Data on the Effectiveness of Psychotherapy*

unadjusted variance for the normal approximation of the test, with an adjustment for ties, and the resulting adjusted variance. SQ then lists the null hypothesis and gives the results of running the normal approximation of the test.

Here we can see that we would easily reject the null hypothesis that the two medians are equal, with the probability of obtaining these results by chance of less than 0.01.

# Kruskal-Wallis Equality of Populations Test

The Kruskal-Wallis one-way equality of populations test is used to test the hypothesis that several samples are from the same population. The test assumes that each sample is random, the observations are independent, the variable being studied is continuous and at least ordinal (ranks), and that the populations being sampled from are the same except that at least one has a difference in location (Daniel, 1978, 201). Each subgroup should have at least five observations, and the subgroups (samples) do not have to be equally sized.

This test is a nonparametric test for a one-way analysis of variance (using ranks instead of the original data). You obtain the same test by choosing *ANOVA* and then *One-way nonparametric* from the *Statistics* menu.

The null hypothesis is that the medians of each subpopulation are identical. The alternative hypothesis is that not every subpopulation has the same median. Analysis of variance tests a similar hypothesis about means instead of medians.

The test works with ranks instead of raw data, comparing the sums of ranks in each sub-population. When each subsample has at least 5 observations and there are at least 3 subsamples, the test statistic is distributed as a chi-squared statistic with k-1 degrees of freedom. StataQuest runs the chi-squared statistic and reports the degrees of freedom and significance level.

The file *lowbw.dta* contains 189 cases from a study of the factors affecting birth weights. We shall consider whether there is a significant difference in birth weight (variable *bwt*) in two groups, those who had mothers who smoked and those who had nonsmoking mothers (variable *smoke*).

To run the Kruskal-Wallis test:

**1** From the *File* menu, choose *Open*.

**2** Enter the name of the file: lowbw.

**3** From the *Statistics* menu, choose *Nonparametric tests* and then *Kruskal-Wallis*.

**4** **Windows:** The Kruskal-Wallis dialog box asks for "Data Variable" and "Group Var." Click on *bwt* for the data variable and *smoke* for the group variable. Click on OK.

   **DOS:** In answer to the questions, the variable to be tested should be bwt and the group variable should be smoke.

**5** We then repeat the analysis to ascertain whether there are similar differences between those mothers with a history of hypertension (variable *ht*) and those without.

The results are shown in Figure 14.5.

The test statistic is a chi-squared statistic with degrees of freedom equal to the number of groups minus one. In the first case, there is clearly a significant difference in baby's birth weights between the mothers who smoke and those who do not, that is, our result is unlikely to have arisen by chance from a population where both groups are identical. In the second case, comparing mothers who have a history of hypertension (high blood pressure) with those who do not, the difference could have arisen by chance 0.1163 of the time.

```
  Stata Results

  . kwallis bwt, by(smoke)

  Test: Equality of populations (Kruskal-Wallis Test)

        smoke        _Obs     _RankSum
           No         115     11913.50
          Yes          74      6041.50

  chi-squared =      7.252 with 1 d.f.
  probability =      0.0071

  . kwallis bwt, by(ht)

  Test: Equality of populations (Kruskal-Wallis Test)

           ht        _Obs     _RankSum
            0         177     17103.00
            1          12       852.00

  chi-squared =      2.466 with 1 d.f.
  probability =      0.1163
```

**Figure 14.5** *The Kruskal-Wallis Test Applied to Data From a Study of Low Birth Weights and Maternal Characteristics*

# Kolmogorov-Smirnov Tests

StataQuest will run either the single-sample or the two-sample Kolmogorov-Smirnov (KS) test. The single-sample test runs only in command mode or through a Command window. The two-sample test runs through SQ's menu system.

## Single-Sample Kolmogorov-Smirnov Test

The single sample KS test can be run through command mode or through a Command window in SQ for Windows. Enter the command help ksmirnov for directions.

## Two-Sample Kolmogorov-Smirnov Equality of Distributions Test

The two-sample version is used to compare two sample distributions, similar to the circumstances when the Mann-Whitney and Kruskal-Wallis tests are used. The null hypothesis is that the two distributions are the same. In general, the KS test is more powerful than the Kruskal-Wallis test, and there is no requirement of a minimum number of observations in either category. The test requires at least ordinal level data. The StataQuest implementation of the KS test requires that the data be arranged so that there is a single data variable, along with a group variable dividing the data into two categories, which are compared to each other.

To run the Kolmogorov-Smirnov two-sample test:

**1** From the *File* menu, choose *Open*.

**2** Enter the name of the file: lowbw.

**3** From the *Statistics* menu, choose *Nonparametric tests* and then *Kolmogorov-Smirnov*.

**4** **Windows:** The Kolmogorov-Smirnov dialog box asks for "Data Variable " and "Group Var." Click on *bwt* for the data variable and *ht* for the group variable. Click on OK.

**DOS:** In answer to the questions, the variable to be tested should be bwt and the group variable should be ht.

The result is shown in Figure 14.6.

## Stata Results

```
. ksmirnov bwt, by(ht)

Two-sample Kolomogorov-Smirnov test for equality of
distribution functions

  Smaller group        D      P-value  Corrected
  -------------------------------------------------
  0:                 0.0466    0.952
  1:                -0.2994    0.133
  Combined K-S:      0.2994    0.266      0.176
```

**Figure 14.6** *Kolmogorov-Smirnov Two-Sample Test*

The first line of the results (under the words *Smaller group*) in Figure 14.6 tests the hypothesis that the birth weight for the first group (nonhypertensive) contains smaller values than the second group (the 7% of the sample who have a history of hypertension). The p-value of 0.952 indicates that we have a high probability of having obtained this result by chance. The second line tests the hypothesis that the birth weights for the nonhypertensive group are larger than those in the hypertensive group (a history of hypertension is coded 1 on the *ht* variable). Although the probability that the result could have arisen by chance is much lower, it is still not below the 0.05 that is the convention for rejecting a null hypothesis. The third line is the actual KS test, and it too has a probability that is high enough at 0.266 that we would normally conclude that the results arose by chance and that there is no significant difference between the two groups on birthweight.

The **corrected** probability value in the last column is a less conservative approximation for the conservative tables developed by Smirnov in 1939 (see SRM, 1993, II, 5s, ksmirnov).

## Summary

The following chart summarizes some information about each nonparametric test included in SQ:

| Test | No. of variables | Data | SQ features |
|------|------------------|------|-------------|
| **Sign Test** | • 2 variables<br>• matched pairs of data | • ordinal & continuous data<br>• random sample<br>• each obs. independent | SQ runs binomial test |
| **Wilcoxon Signed-Ranks Test** | • 2 variables<br>• matched pairs of data | • interval level data<br>• n > 25, each sample<br>• same n, each sample | SQ runs large sample approximation with Z score |
| **Mann-Whitney Two Sample Test** | • 2 variables<br>• 1 data variable<br>• 1 group variable | • 2 random, independent samples<br>• data variables continuous and ordinal or better | SQ runs normal approximation with Z score |
| **Kruskal-Wallis Equality of Distribution Test** | • 2+ samples from same population<br>• 1 data variable<br>• 1 group variable | • random, independent samples<br>• data variables continuous and ordinal or better | • Nonparametric test for ANOVA<br>• SQ runs chi-squared with k-1 degrees of freedom |
| **Kolmogorov-Smirnov Two-Sample Equality of Distribution Test** | • 2 samples or variables<br>• 1 data variable<br>• 1 group variable | • ordinal level data or better<br>• 2 independent, random samples | |

## ▨ EXERCISES

**14.1** Use the *paulch14.dta* data set to run the Sign test, the Wilcoxon signed-ranks test, and the paired t-test on the two sets of variables, *timet1/timec1*, and *timet2/timec2*. How different are the results? Why?

**14.2** The general fertility rate is the ratio of the number of live births per year to the total number of females aged 15 to 44. The total fertility rate, also included in the data set *fertilty.dta*, is "the number of births 1,000 women ages 10 to 50 would have in their lifetime if at each year of age they experienced the birth rates occurring to women of that age in the specified calendar year" (Wright, 1989, 240). Test whether the variable for the general fertility rate (*genlrate*) is significantly different during the early boom years, 1946–65, compared with the years before and after. The group variable is *boomyr*.

**14.3** Use the *pauling.dta* version of the Vitamin C / terminal cancer patient data to use the Mann-Whitney test to test a similar hypothesis to that tested in Figure 14.3. *pauling.dta* has the same data in it as *paulch14.dta*; the difference is how the data are arranged. The relevant variables are *a* (includes the data for both treatment and control groups from variables *timet1* and *timec1*) and *b* (includes both *timet2* and *timec2*).

**14.4** Use the data set *husbands.dta* to test the hypothesis that the husbands in the study had the same ages as their wives. Use both the parametric paired t-test and the nonparametric Wilcoxon signed-ranks test to gather evidence concerning the hypothesis. Explain the results.

**14.5** Use the same data set as in problem 14.4 to test the hypothesis that husbands have the same heights as their wives. Explain your results.

**14.6** Using the data set referred to in the section of this chapter on the Sign test, *autoexp1.dta*, use a paired t-test, the Sign test, and the Wilcoxon signed-ranks test to determine whether there is a significant difference before and after the experiment. Explain your results.

**14.7** The data set *deaths.dta* contains a data set of 1,251 famous men and when they died in relation to their birthdays. Use the Mann-Whitney test to decide whether or not famous men usually die just after, instead of just before their birthmonths.

**14.8** The data set *birds.dta* has several items of information on 32 New Zealand birds. Use the Mann-Whitney test to ascertain whether the birds that fly are different from those that are flightless in their body mass, basal rate, and pectoralis (variables *bodymass*, *basalrat*, and *pectopct*).

**14.9** Use the Kruskal-Wallis test to ascertain whether there are significant differences in the variable for systolic blood pressure among the several different drugs tried in the experiment in *sysage.dta*, and among the several different diseases. The data variable is *systolic*; group variables are *drug* and *disease*.

**14.10** Use the appropriate nonparametric test to ascertain whether there is a significant difference in birth weights (*birthwt*) for infants with severe idiopathic distress syndrome between those who lived and those who died (*risk*). The data set is *respdist.dta*. Compare your results with the parametric test in problem 13.3.

**14.11** In problem 13.1, you used a two-sample t-test to ascertain whether there was a significant difference between British Kings and Queens in their lifespans after taking office as compared with U.S. Presidents and Roman Catholic Popes. Now use the Mann-Whitney test on the *kings* variable and the Kruskal-Wallis test on the *leader* variable. Explain the difference in your results. The data set is *pres.dta*.

**14.12** Use the Kruskal-Wallis test to ascertain whether there is a significant difference among the three groups of college students in *caffeine.dta*. The data variable is *taps*; group variable is *group1*.

**14.13** The data set *crying.dta* contains the result of an experiment on the effects of rocking babies on their incidence of crying in a hospital. Use the Sign test and the Wilcoxon signed-ranks test to ascertain whether there is a difference in the percentage of the babies crying each day. The relevant variables are *ctrlsucc* and *rocksucc*.

# 15

# *The Statistical Calculator*

## What's a Statistical Calculator?

A **calculator** allows you to enter numbers from a keypad and then perform mathematical tasks on them. A **statistical calculator** allows you to enter summary data from the keyboard and then obtain statistical results. For example, you can run a one-sample t-test by entering the sample size, the sample mean, the sample standard deviation, and the hypothesized mean. StataQuest will produce a t-value and the probability of having obtained that t-value by chance.

The standard calculator is the second-to-last menu item. Each command in this chapter executes when you choose Run (SQ for Windows) or when you finish answering the questions (SQ for DOS).

## One-Sample Normal Test

The one-sample normal test is used to obtain the z-value and significance level of a sample mean compared with an assumed population mean. It is assumed that the population variance is known, a requirement that limits the use of the test. If the population variance is not known, you should use a one-sample t-test, where the sample variance estimates the population variance.

■ ■ ■ ■ ■ ■ ■

# IN THIS CHAPTER

- ■ **One-sample normal test**
- ■ **Two-sample normal test**
- ■ **One-sample t-test**
- ■ **Two-sample t-test**
- ■ **One-sample test of proportion**
- ■ **Two-sample test of proportion**
- ■ **One-sample and two-sample test of variance**
- ■ **Confidence interval for the mean**
- ■ **Binomial confidence interval**
- ■ **Poisson confidence interval**
- ■ **Statistical tables (normal, Student's-t, chi-squared, binomial, Poisson)**
- ■ **Inverse statistical tables (normal, Student's-t, F, chi-squared)**
- ■ **Standard calculator**
- ■ **RPN calculator**
- ■ **Checking a cross-tab printed in a book**

Devore and Peck (1993, 436–7) present a sample problem. The National Academy of Sciences has recommended that the average daily sodium intake should not exceed 3,300 mg. *Consumer Reports*, however, found from a sample of 100 random Americans that the average sodium intake for that group was 4,600. The population standard deviation is 1,100 mg. The question is whether the mean intake for our sample is significantly different from the recommended average of 3,300.

**1**   From the *Calculator* menu, choose *1-sample normal test.*

**2**   **Data to be input:**

Number of observations: 100

Sample mean: 4,600

Population standard deviation: 1,100

Hypothesized mean: 3,300

```
╔═══════════════════════════════════════════════════════════════╗
║  Stata Results                                                 ║
╠═══════════════════════════════════════════════════════════════╣
║                                                               ║
║  . ztesti 4600 1100 3300 100, level (95) z = 0.12             ║
║                                                               ║
║     Variable |      Obs        Mean    Std. Dev.             ║
║   -----------+-------------------------------------           ║
║           x  |      100        4600         1100             ║
║                                                               ║
║          Ho:   mean  = 3300                                  ║
║                   z  = 11.82                                 ║
║          Pr > |z|  = 0.0000                                  ║
║             95% CI  = (4419.066,4780.934)                    ║
║                                                               ║
╚═══════════════════════════════════════════════════════════════╝
```

**Figure 15.1** *One-Sample Normal Test*

The results are shown in Figure 15.1. We are able to reject the null hypothesis.

Circumstances where the population standard deviation will be known are, for example, where a survey is done just after the decennial Census of Population and Housing and the Census has provided accurate data on the population standard deviation, or with some standardized tests. For example, the Scholastic Aptitude Test has been developed and standardized with a goal of having a mean of 500 and standard deviation of 100.

## Two-Sample Normal Test

The two-sample normal test is used to test the null hypothesis that there is no difference between two sample means in a situation where the population variances are both known. The two-sample t-test is used when the variances are unknown and must be estimated from the sample variances. StataQuest lists the null hypothesis "Ho: mean(X) = mean(Y)."

Continuing with the problem from the previous section, let us suppose that the National Academy of Sciences has now found on the basis of a sample of 500 respondents that the average daily sodium intake for Californians is 3,000 mg. A national survey of 1,000 dieters has found that they average 2,800 mg of sodium. The population standard deviation is 1,100 mg. The question is whether the two groups could differ only randomly.

**1**   From the *Calculator* menu, choose *2-sample normal test.*

**2**   **Data to be input:**

Number of observations: 500 for Sample 1 and 1,000 for Sample 2

Sample mean: 3,000 for Sample 1 and 2,800 for Sample 2

Population standard deviation: 1,100 for both samples

The results are shown in Figure 15.2.

---

**Stata Results**

```
. ztest2i 500 3000 1100 1000 2800 1100, level(95)

Variable |      Obs       Mean     Std. Dev.
---------+-----------------------------------
       x |      500       3000       1100
       y |     1000       2800       1100
---------+-----------------------------------
combined |     1500    2866.667    60.24948

         Ho: mean(x) = mean(y)
                z =  3.3195307
          Pr > |z| =  0.0009
            95% CI = (81.913,318.087)
```

---

**Figure 15.2** *Two-Sample Normal Test*

Figure 15.2 shows a very low probability (0.0009) that these results could have arisen by chance, and you decide to reject the null hypothesis.

## One-Sample t-Test

The one-sample t-test is used to obtain the t-value and significance level of a sample mean compared with an assumed population mean. The significance level will yield the probability that results this extreme or more could have arisen by chance under the null hypothesis that the sample mean equals the assumed population mean. If the probability is low that these results arose by chance, then there is little likelihood that the sample was drawn from a population where the null hypothesis is true.

Consider the Graduate Record Exam, where the goal is to have a verbal score mean of 500 and standard deviation of 100. The mean verbal score for 2,000 seniors in Political Science on one particular testing date was 495, with a standard deviation of 115. Is the test performing as planned, based on this particular sample?

1 From the *Calculator* menu, choose *1-sample t test*.

2 **Data to be input:**

Number of observations: 2000

Sample mean: 495

Sample standard deviation: 115

Hypothesized mean: 500

The results are shown in Figure 15.3.

**Stata Results**

```
. ttesti 2000 495 115 500

Variable |      Obs        Mean    Std. Dev.
---------+-------------------------------
       x |     2000         495         115

       Ho:   mean = 500
                t = -1.94 with 1999 d.f.
       Pr > |t| = 0.0520
          95% CI = (489.95694,500.04306)
```

**Figure 15.3** *One-Sample t-Test*

In Figure 15.3, we find that the probability of obtaining these results by chance is 0.052, and we decide on the basis of this relatively high probability not to reject the null hypothesis. The sample with a mean of 495 and a standard deviation of 115 could well have been drawn from a population with a mean of 500.

The degrees of freedom for this test are the sample size minus 1 (n–1), which is 1,999 in Figure 15.3.

**Assumptions:** Again, we are assuming that the underlying population distribution is normally distributed. The true population variance is unknown and is estimated by the sample variance. If the true population variance is known, then a z-value tested against the normal distribution can be used (see Kanji, 1993, 21).

## Two-Sample t-Test

The two-sample t-test is a test of the significance of the difference between the means of two populations. Both means are from samples drawn randomly from the two underlying populations. We are assuming that the underlying populations are distributed normally, and that the true population variances are unknown. We are actually testing whether the difference between the two means is equal to zero, which is the same as testing whether one mean is equal to the second one.

Consider two groups of students who took the Graduate Record Examinations. The Sociology group nationally was composed of 1,000 students who averaged 512 on the verbal portion of the exam. The sample standard deviation is 95. The Political Science group comprises 1,500 students who averaged 495 on the exam, with a sample standard deviation of 115. Is the difference between 495 and 512 statistically significant?

1 From the *Calculator* menu, choose *2-sample normal test*.

2 **Data to be input:**

Number of observations: 1,000 for Sample 1 and 1,500 for Sample 2 (remember not to input the comma).

Sample mean: 512 for Sample 1 and 495 for Sample 2.

Sample standard deviation: 95 for Sample 1 and 115 for sample 2.

3 **Windows:** Do not check the box for "unequal variances" as yet.

**DOS:** Answer yes to the question about equal variances.

The results are shown in Figure 15.4.

---

## Stata Results

```
. ttesti 1000 512 95 1500 495 115

    Variable |      Obs         Mean    Std. Dev.
    ---------+-----------------------------------
           x |     1000          512           95
           y |     1500          495          115
    ---------+-----------------------------------
    combined |     2500        501.8     107.7502

Ho:  mean(x) - mean(y) = 0   (assuming equal variances)
                  t = 3.88 with 2498 d.f.
            Pr > |t| = 0.0001
             95% CI = (8.39826,25.60174)
```

**Figure 15.4** *Two-Sample t-Test*

---

The results show that the null hypothesis should be rejected at less than 0.01.

One further choice not present in the previous tests is whether the variances in the two samples are equal. If the variances can be assumed to be equal, then the test yields a slightly higher t-value and lower probability that the results could have arisen by chance. You can check the box and see what difference this option makes in this problem.

## One-Sample Test of Proportion

The one-sample test of proportion is used when the underlying process can be classified into either of two categories. These can be success or failure, heads or tails, having or not having a disease, or whatever. These are binomial processes, and they are also independent; the probability of success in any one trial is unaffected by previous trials.

Consider students taking a test where you know the results for past occasions when you have given the test. You have 50 students, and 35 pass the test. Your previous research, however, indicates that only 50% should have passed the test. Do you have a particularly good class compared with those who have taken the test previously? The binomial probability test can tell you how likely it is that these results came from a population where the population proportion is .50.

**1** From the *Calculator* menu, choose *1-sample test of proportion.*

**2** **Data to be input:**

Number of observations: 50 (for the number of students in our current sample)

"No. of successes" Here we input our sample results. In our 50-person sample, we had 35 who passed the test. So we input: 35.

"Prob. of success or exp. no. of successes"—here SQ is looking for our past results. We can input this number either in the form of a "probability of success," that is, any number greater than 0 and less than 1, or we can input the actual number of successes we expected. So we could input either .5 (in the past 50% have passed the test) or 25, the number of students who "should" have passed the test based on previous results.

**3** **Windows:** Do not check the box for "normal approximation" as yet. Check Run.

**DOS:** Answer yes to the question about the normal approximation.

The results are shown in Figure 15.5.

The output in Figure 15.5 is one of the most complex in StataQuest. First, note that SQ as usual tells us the command in command language: ". `bintesti 50 35 .5`". Then:

***Varname* - x.** StataQuest makes up a variable name because we have no data set in memory.

***Obs. N* - 50.** StataQuest tells us the number of observations we inputted.

```
┌─────────────────────────────────────────────────────────────────────┐
│ Stata Results                                                         │
├─────────────────────────────────────────────────────────────────────┤
│ . bintesti 50 35 .5                                                   │
│                                                                       │
│    Varname |   Obs. N    Obs. k      Exp. k      Assumed p    Obs. p  │
│ -----------+---------------------------------------------------------  │
│          x |       50        35          25       0.50000    0.70000  │
│                                                                       │
│       [1]  Pr(k>=35)              = 0.003300   (one-sided test)       │
│       [2]  Pr(k<=35)              = 0.998699                          │
│       [3]  2*min([1],[2])         = 0.006600   (two-sided test)       │
│                                                                       │
│       [4]  Pr(k==35)              = 0.001999                          │
│       [5]  Pr(k==16)              = 0.004373                          │
│       [6]  Pr(k==15)              = 0.001999                          │
│       [7]  Pr(k>=35 | k<=15)      = 0.006600   (two-sided test)       │
│                                                                       │
│ Ho: proportion = .5                                                   │
│                                                   -- Binomial Exact -- │
│ Variable |    Obs      Mean     Std. Err.      [95% Conf. Interval]    │
│ ----------------------------------------------------------------------  │
│          |     50       .7      .0648074       .5537109    .8212891    │
└─────────────────────────────────────────────────────────────────────┘
```

**Figure 15.5** *One-Sample Test of Proportion (Binomial Probability Test)*

*Obs. k - 35.* k is the conventional binomial symbol for the number of successes. Notice that this is the actual number. We observed 35 successes; we expected 25.

*Exp. k - 25.* This is the number of successes we should have had in our observations if the population probability (the hypothesized probability of success) were true. This number is the number of observations times the hypothesized probability of success = 50 x .50 = 25.

*Assumed p - 0.50000.* This is the hypothesized probability of success that we inputted.

*Obs. p - 0.70000.* This is the probability we actually observed in our sample, arrived at by dividing the *Obs. k* by the *Obs. N* = k/N = 35/50 = 0.70. StataQuest prints out this result with five places to the right of the decimal point because the binomial test is sometimes used with medical problems dealing with the incidence of a disease where we might have 10 observed cases in a population of several million; thus the need for several places to the right of the decimal point.

The output can then be interpreted as follows:

**Line 1.** A one-tailed test of the probability of observing 35 or *more* success-es. This is a very unlikely probability (0.003300). It is not likely, given that the rate for previous students was 0.50 or 25 successes, that we got this result by chance. Either Line 1 or Line 2 is labeled as a "(one-tailed test)".

**Line 2.** The probability of getting 35 or *fewer* successes. If our observed suc-cesses had been less than 25 (the expected number of successes), then Line 2 would have been labeled as the (*one-tailed test*).

**Line 3.** A crude two-tailed test that is twice the one-tailed test on Line 1 or 2. Here it is twice Line 1. A more sophisticated two-tailed test is derived in Lines 4 through 6 and contained on Line 7.

**Line 4.** The probability of exactly k (here 35) number of successes.

**Line 5 and 6.** Part of the process of deriving the two-tailed test probability. Here we have 35 successes on one side of the distribution. What would be the corresponding levels on the other side of the distribu-tion? StataQuest looks at both 16 successes and 15 successes and prints them in Lines 5 and 6.

**Line 7.** The actual two-tailed test probability. Only if the assumed probabil-ity of success in the population is 0.50 (which it is in our example) will Lines 3 and 7 be the same.

Result: The probability of obtaining our result by chance is very low, 0.0033 on a one-tailed test or 0.0066 on a two-tailed test. Conceptually, the one-tailed test (of obtaining 35 or more successes) makes much more sense in our case.

## Normal Approximation (Windows Only)

Many textbooks also present a z-test that approximates the binomial distrib-ution when the number of observations is "large," that is, more than 30. It can be obtained by following the previous directions and checking the box next to "Normal Approximation." The results for the previous problem are in Figure 15.6.

In Figure 15.6 we have a probability of obtaining these results or more extreme results of 0.0047, which is at least "in the same ball park" as the results obtained with the binomial test itself.

## Another Example

Figure 15.7 shows a medical case involving much larger numbers. We have found 15 cases of X disease in a given city of 3,000,000 population over a peri-od when we would expect the rate of disease for this population to be 0.00001.

```
Stata Results

. bintesti 50 35 .5 , normal

    Variable |      Obs     Proportion   Std. Error
    ---------+---------------------------------------
           x |       50            35      .0648074

           Ho:   P = .5
                 z = 2.83
         Pr > |z| = 0.0047
            95% CI = (.583691,.816309)
```

**Figure 15.6** *Normal Approximation to the One-Sample Test of Proportion*

```
Stata Results

. bintesti 3000000 15 .00001 , normal

    Variable |      Obs     Proportion   Std. Error
    ---------+---------------------------------------
           x |   3.0e+06      5.00e-06     1.29e-06

           Ho:   P = .00001
                 z = -2.74
         Pr > |z| = 0.0062
            95% CI = (2.000e-06,8.000e-06)
```

**Figure 15.7** *Normal Approximation to the One-Sample Test of Proportion*

We have run the normal approximation in Figure 15.7. Running and interpreting the binomial test is left as an exercise for the reader. There is a very low probability of obtaining results of 15 cases or less from a population with an incidence of the disease of 0.00001. We should have had 3,000,000 x .00001 = 30 cases during this time period. For instructions on interpreting scientific notation (2.000e–06), see page 218.

## Two-Sample Test of Proportion

The two-sample test of proportion is a test of the significance of the difference between proportions drawn from two different populations. The test is a normal approximation that assumes that the sample sizes from each population are "large," that is, more than approximately 30. Both proportions are from samples drawn randomly from the two underlying populations.

Consider a political candidate engaged in a tight race. Last week's poll put the candidate's support at 45%, based on a random sample of 100 drawn from the lists of registered voters. This week's poll, again based on 100 randomly selected voters, has the candidate at 55%. Is jubilation justified?

**1** From the *Calculator* menu, choose *2-sample test of proportions*.

**2** **Data to be input:**

Number of observations: 100 for Sample 1 and 100 for Sample 2

No. of successes or observed proportion: .55 for Sample 1 and .45 for Sample 2.

The results are shown in Figure 15.8.

---

**Stata Results**

```
. prtesti 100 .55 100 .45, level(95)

   Variable |      Obs    Proportion
  ----------+----------------------
          x |      100          .55
          y |      100          .45
  ----------+----------------------
   combined |      200           .5

      Ho:  proportion(x) - proportion(y) = 0
                                      z = 1.41
                             Pr > |z| = 0.1573
                  95% conf. interval = (-0.0379, 0.2379)
```

---

**Figure 15.8** *Two-Sample Test of Proportions*

The results in Figure 15.8 show that there is a high probability, 0.1573, that these results could have arisen by chance and that there may be no difference between the two sample proportions, that is, they could both have been drawn from the same underlying population. The 95% confidence interval of the difference between the two proportions (the difference is .55-.45=.10) is from a lower value of -0.022 to an upper value of 0.222, an interval that includes 0, that is, there may be zero difference between the two sample proportions.

The candidate concludes that the sample sizes are too small. After consulting with a student intern who just concluded a statistics course, she decides to increase the sample size to 400. The results are up again, to 57%.

Is there now a significant difference between 100 observations and .45 and 400 observations and .57? (The analysis is left to the reader; the probability of obtaining a z-value of this size or larger should be found to be 0.0311. What would you conclude on the basis of the probability?)

## One-Sample and Two-Sample Test of Variance

The one-sample test of variance tests the hypothesis that a given standard deviation (the square root of the variance) is equal to a number you supply. The result is a chi-squared statistic and the probability that it could have arisen by chance if the null hypothesis were true.

Given a difference between two sample standard deviations, the two-sample test of variance performs an F test and computes the probability that a difference between the two sample standard deviations that large or larger could have been obtained from a population under the null hypothesis that the difference between the two standard deviations is zero. This test is also called a variance ratio test because it is based on the ratio of the two variances.

We shall test whether a sample of 100 cases with a standard deviation of 6 has the same standard deviation as a second sample of 100 with a standard deviation of 5.1.

**1**  From the *Calculator* menu, choose *2-sample test of of variance*.
**2**  **Data to be input:**
  Number of observations: 100 for Sample 1 and 100 for Sample 2
  Sample standard deviation: 6 for Sample 1 and 5.1 for Sample 2

The results are shown in Figure 15.9.

Figure 15.9 shows an F value, which is 1.38. Like all F values, it has two degrees of freedom, which are the number of observations in the first sample minus one and the number of observations in the second sample minus one, or 99 and 99. The probability that this value could have arisen by chance is more than 10% at 0.1075, which is generally considered too high to reject the null hypothesis. We conclude that the evidence is not sufficient to reject the null hypothesis that the two sample standard deviations could have come from the same population, or, equivalently, that the difference between the two sample standard deviations (reformulating the null hypothesis to the equivalent "Ho: sd(x) - sd(y) = 0") is zero.

## Confidence Interval for the Mean

A confidence interval is an interval, derived from our sample, within which we expect, with a certain probability level, to find the mean of the population we are

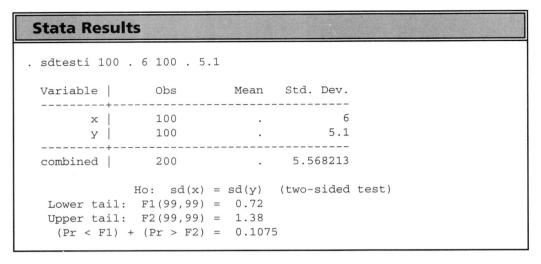

```
Stata Results

. sdtesti 100 . 6 100 . 5.1

Variable |      Obs        Mean      Std. Dev.
---------+-----------------------------------
       x |      100          .              6
       y |      100          .            5.1
---------+-----------------------------------
combined |      200          .       5.568213

              Ho:  sd(x) = sd(y)   (two-sided test)
  Lower tail:  F1(99,99) =   0.72
  Upper tail:  F2(99,99) =   1.38
    (Pr < F1) + (Pr > F2) =  0.1075
```

**Figure 15.9** *Two-Sample Test of Variance*

making inferences about. A confidence interval is "an estimated range of values with a given probability of covering the true population mean" (Hays and Winkler, 1971, 327). If the level of confidence is 95%, then, if we were to take a large number of samples and compute confidence intervals for each, approximately 95% of these confidence intervals would include the true population mean.

Hamilton (1990, 263) notes a European study that 44 different brands of cigarettes had a mean cadmium content or 1.4 mg/g, with a standard deviation of .4. If we assume that the 44 brands are a random sample of all cigarette brands, then we would construct a 95% confidence interval around the 1.4 mg/g figure as follows:

**1**   From the *Calculator* menu, choose *Confidence interval for mean*.

**2**   **Data to be input:**

Number of observations: 44

Sample mean: 1.4

Sample standard deviation: .4

Confidence level: the level is given at "95," but we can change it if we want. In this case we do not do so.

The results are shown in Figure 15.10.

We decide to re-run the analysis and formulate a 99% confidence interval. To do so, we choose the same dialog box and change only the confidence level from 95 to 99. The results are shown in the bottom half of Figure 15.10.

```
. cii 44 1.4 .4, level(95)

Variable |      Obs      Mean    Std. Err.     [95% Conf. Interval]
---------+--------------------------------------------------------
         |       44       1.4    .0603023      1.278389    1.521611

. cii 44 1.4 .4, level(99)

Variable |      Obs      Mean    Std. Err.     [99% Conf. Interval]
---------+--------------------------------------------------------
         |       44       1.4    .0603023      1.237479    1.562521
```

**Figure 15.10** *Confidence Interval for the Mean*

In Figure 15.10, StataQuest tells us that the 95% confidence interval is from 1.28 to 1.52. StataQuest also gives us the number of observations (Obs), the Mean (1.4), and the standard error (*Std. Err.*). The standard error, of course, follows the familiar formula of the sample standard deviation divided by the square root of the sample size. The 99% confidence interval is wider, from 1.24 to 1.56.

**Assumptions:** Doing a confidence interval assumes that the underlying population distribution from which the sample is drawn is distributed normally and that the sample is random.

# Binomial Confidence Interval

A binomial confidence interval assumes that we have a process that is dichotomous (success/failure, and so on) and that we want a confidence interval around the number of successes. The interval depends upon the number of successes, the sample size, and the level of confidence. We shall run a confidence interval around the 35 "successes" out of 50 students who took the test referred to in an earlier section of this chapter.

**1**  From the *Calculator* menu, choose *Binomial confidence interval*.
**2**  **Data to be input:**
   Number of observations: 50 [sample size]
   Number of successes: 35
   Confidence level: 95.

The results are shown in Figure 15.11.

```
 Stata Results

 . cii 50 35, level(95)
                                                   -- Binomial Exact --
 Variable |     Obs      Mean     Std. Err.      [95% Conf. Interval]
 ---------+-----------------------------------------------------------
          |      50        .7     .0648074        .5537109    .8212891
```

**Figure 15.11** *Binomial Confidence Interval, N=50, k=35*

In Figure 15.11 we have a confidence interval from 0.55 to 0.82. However, there is a special interpretation of binomial confidence intervals, which goes as follows: If the true probability of passing the test were 0.55, the probability of obtaining a result as extreme or more extreme than 35 or fewer passing would be 2.5% (half the 5% from the 95% confidence interval). If the true probability of passing the test were 0.82, the probability of obtaining a result as extreme or more extreme than what we obtained (35 or more passing) would be 2.5%.

Notice that the confidence interval does not include 0.50, the rate at which previous students had passed the exam.

We shall run a similar confidence interval for the medical problem, where we had a city of 3,000,000 with 15 cases of a disease. The result is shown in Figure 15.12.

```
 Stata Results

 . cii 3000000 15, level(95)
                                                   -- Binomial Exact --
 Variable |     Obs      Mean     Std. Err.      [95% Conf. Interval]
 ---------+-----------------------------------------------------------
          | 3000000   5.00e-06    1.29e-06       2.80e-06    8.25e-06
```

**Figure 15.12** *Binomial Confidence Interval, N=3,000,000, k=15*

What has happened? StataQuest has switched to scientific notation because the numbers were too big or too small to print in the space available. To convert these numbers to regular notation, move the decimal point the required number of spaces to the right or to the left. Thus 5.00e-06 requires the decimal point to be moved six spaces to the left, and the number is 0.000005. 1.29e-06 is 0.00000129. Our interval runs from 0.0000028 to 0.00000825 and has an interpretation similar to the earlier interpretation.

# Poisson Confidence Interval

In a Poisson process, we have a situation where there are events that occur over a certain time period. We do not know the total number of events, but we know how many took place over the given time period. We want a confidence interval around the number of events. Our example is the number of traffic accidents that took place in a city in a 24-hour period. There were 163.

**1**   From the *Calculator* menu, choose *Poisson confidence interval.*

**2**   **Data to be input:**

Total exposure time: 1

Number of events: 163

Confidence level: the level is given at "95," but we can change it if we want. In this case we do not do so.

The results are shown in Figure 15.13.

## Stata Results

```
. cii 1 163, poisson level(95)

                                        -- Poisson  Exact --
Variable |  Exposure     Mean    Std. Err.    [95% Conf. Interval]
---------+--------------------------------------------------------
         |        1        163   12.76715      138.9371    190.0326
```

**Figure 15.13** *Poisson Confidence Interval*

Figure 15.12 shows that the 95% confidence interval runs from 138.9 to 190.0 traffic accidents per 24-hour period.

# Statistical Tables

StataQuest's *Statistical tables* submenu enables you to look up the probability in the tails of the distribution of the following distributions: normal, Student's-t, F, chi-squared, binomial, and Poisson. For each, you select the distribution you want, and then one or more items of information, depending upon what parameters define the particular distribution. For each distribution, we discuss what information you enter, and how you interpret the results. To save space, we have not reprinted the results of each action; you are invited to run each example as it is discussed.

## Normal

With the normal distribution, you input a z-value, which is the number of units of standard deviation or standard error from the mean. StataQuest replies with the area under the tail of the distribution. StataQuest will give the area to the left of a one-tailed test, the area of the tail, and the area of both tails for a two-sided test. When StataQuest asks for a *Normal statistic*, it is asking for a z-value.

For our example, we inputted a z-value of 1.96; the output indicates that the area under the tail of the distribution is .025 for a one-tail test and 0.05 for a two-tail test.

## Student's-t

For the t-distribution, you input both the number of degrees of freedom and the actual t-statistic. The number of degrees of freedom for the t-distribution in many tests is the sample size minus 1 (n-1). For our example, we inputted the same t-value as our z-value above (1.96) because the t-distribution is noted for its slightly thicker tails. Our sample size is 10; thus our degrees of freedom are nine. We found the output indicates that the area under the tail of the distribution is .0408 for a one-tail test and 0.0816 for a two-tail test, slightly higher (as it should be) than the normal distribution above.

## F

For the F distribution, you enter the number of degrees of freedom for both the numerator and the denominator of the statistic as well as the F value for which you want the probability. You receive the probability under the right-hand tail of the distribution. We inputted a F value of 1.96, a numerator degrees of freedom of 32, and a denominator degrees of freedom of 164, and obtained the area under the right tail, a "right-tail probability" of 0.0035.

## Chi-Squared

For the chi-squared distribution, you input the number of degrees of freedom and then your actual chi-squared statistic. We find that the probability in the tail of the distribution for a chi-squared statistic of 26.119 with 14 degrees of freedom is 0.025.

## Binomial

For the binomial distribution, you input the number of observations or trials you have observed, the probability of the event that you expected, and the actual successes (which StataQuest has labeled *Observed number of events (k)*). We had 50 observations, a 0.5 probability of success (based on past information), and we observed 35 successes in our sample, the same values as the example for the one-sample test of proportion earlier in this chapter. The three probabilities given here are the probability that k exactly equals 35 (0.0020), the probability that k is greater than or equal to 35 (0.0033), and the probability that k is less than or equal to 35 (0.9987). These correspond to lines 4, 1, and 2 respectively of the one-sample test of proportion (See Figure 15.5). Notice that the probabilities here, however, have been rounded to four digits. With history telling us that we would expect 0.5 of the students to pass, we expected 25 students to pass, but got 35 students with passing scores. Thus, the probability that k is greater than or equal to 35 (0.0033) represents the probability in the tail of the binomial distribution for a one-tail test.

### Poisson

For the Poisson distribution, you input lambda, the expected number of events (the number of trials times the probability of an occurrence). You then input the *Observed number of events (k)*. These two parameters define the distribution. In our example, we expected 138 events (lambda), but observed (k) 164. We found that the probability of obtaining exactly 164 successes when we expected 138 is 0.0031. The probability of more than 164 is 0.0169; the probability of less than 164 is 0.9862.

## Inverse Statistical Tables (Windows Only)

StataQuest's *Inverse statistical tables* submenu enables you to input the probability found in one of the two tails of a two-tailed test and get back the appropriate value for the normal, t, F, or chi-squared distribution. For example, if you input the probability 0.025 for the normal distribution, StataQuest responds with the z-value of 1.96, the value of z when 0.025 is in each tail of the distribution.

### Normal

With the normal distribution, you input the probability found in one of the two tails, and you receive back the corresponding z-value.

### Student's-t

With the t-distribution, you input both the probability "$\Pr(T<=t)=$", which is the probability found to the **left** of the t-value you want to obtain, as well as the degrees of freedom. We inputted a probability of 0.025 and degrees of freedom of 20 and obtained in return a t-value of -2.086; thus our probability of 0.025 referred to the probability in the left hand tail of the t-distribution.

## F

With the F distribution, you again are inputting the probability to the left of your actual F value ("Pr(F<=f) = "), as well as the numerator and denominator degrees of freedom. With numerator degrees of freedom of 2 and denominator degrees of freedom of 20, we inputted

- 0.05 and obtained an F value of 0.05143
- 0.95 and obtained an F value of 3.49

Thus in each case, we inputted the probability to the **left** of the F value and received in return the F value itself for that particular probability.

### Chi-Squared

With the chi-squared distribution, you input the probability again to the left of your chi-squared value (**not** what is in the tail of the distribution), with the degrees of freedom, and you obtain the chi-squared critical value. We inputted a probability of 0.90 (so that 0.10 was in the tail), with 10 degrees of freedom, and obtained a chi-squared of 15.987. Our reference manual lists the chi-squared for those parameters (that is, a probability of 0.10) as 15.99.

## Standard Calculator

**Windows:** The Standard Calculator performs the normal calculator functions in response to your input. It works in the normal manner, that is, you input a number, a sign, another number, and equals. Your expression is evaluated, and the result appears in the register at the top. Input "20 + 20 =" and 40 appears at the top. The buttons and keys beyond the normal +, -, * (multiply), and / are the following:

- **1/x** divides the number in the register into 1. If 40 is in the register, pressing **1/x** yields 0.025.
- **log** yields the natural log (to the base e) of whatever number is in the register. If 40 is in the register, pressing **log** yields 3.69. These are logs to the base e (2.71828...).
- **exp** yields the antilog, that is, the opposite of log. If 3.69 is still in our register and we press **exp**, we will be back to 40.
- **+/-** exchanges the sign of the number in the register. If 40 is in the register, pressing **+/-** yields -40.
- **sin** yields the sine of whatever number is in the register. If 40 is in the register, pressing **sin** yields 0.745....
- **cos** yields the cosine of whatever number is in the register. If 40 is in the register, pressing **cos** yields -0.6669....

- **sqrt** yields the square root of whatever number is in the register. If 36 is in the register, pressing **sqrt** yields 6.
- **y^x** means y to the x power. If we input 7, press **y^x**, input 2, and then press =, the register should show 49. Thus 7=y and 2=x.
- **cls** clears the register to 0.

**DOS:** The Expression Evaluator works similarly to a formula; you input the formula to the right of the equals sign. Examples: 20+20; log(10); sqrt(14/42); sqrt(14)/42; (42*300)/(15042+432).

# RPN Calculator (Windows Only)

RPN means Reverse Polish Notation, a scheme developed particularly for engineers and others who must input long and complicated formulas with many mathematical functions and operations. Hewlett-Packard sells several RPN calculators if you want to purchase one. The basic principle is easy:

number ENTER number OPERATION

In other words, instead of 1 + 1 = 2 as in a regular calculator, a RPN calculator works like this: 1 ENTER 1 +. The operation comes AFTER the number it applies to. And there is no equals (=) key.

The RPN calculator is especially convenient (if you are not an engineer) for doing things like balancing checkbooks, where you need to enter a long string of numbers: 33.04 ENTER 20.00 MINUS 12.04 MINUS 150.00 PLUS, and so on. In other words, once you enter the first number plus ENTER, you enter a number and then what happens to it (+ or -) **after** each number.

Another difference with the RPN calculator is that the register at the top is actually a **stack** of numbers. The **swap** button will swap the bottom number for the one just above it in case you wish to reverse operations (instead of dividing 60 by 3, you decide to divide 3 by 60).

The **buttons** discussed previously work the same with the RPN calculator as they do with the regular calculator except for the following:

**All buttons**, when pressed, apply to the number at the bottom of the stack.

**y^x** works like this: 40 ENTER 2 **y^x** yields 1600.

**ENTER** places a number at the bottom of the stack in the register at the top.

**SWAP** swaps the bottom two numbers in the stack.

**DROP** drops the bottom number in the stack and can be used to clear out a mistaken entry so that the number above it will stay in the register.

**CLS** clears out the stack.

Your authors have totally different opinions about the RPN calculator. One of us loves it and uses it to do all his daily calculating; the other thinks it defies rationality. Take your choice.

## Checking a Cross-Tab Printed in a Book

This command enables you to input the number of observations contained in the cells of a cross-tabulation and obtain in return a printed table, the various statistics associated with a table that SQ can print, and the like. If you go to the **Command** window (DOS: *File / Command mode*) and input the word "tabi" followed by the number of observations in the first row, a backslash ("\"), the number of observations in the second row, and so on. Thus, for a 2x2 table with the first row's cells containing 232 and 146 cases respectively, and the second row containing 55 and 335 observations, you would input in the Command window: `tabi 232 146 \ 55 335, all`. Try it.

## ■ EXERCISES

15.1 Test the hypothesis that the population mean is equal to 50 against the alternative hypothesis that it is greater than 50 with the following test data:

a. n = 10, sample mean = 55, sample standard deviation = 10.

b. n = 20, sample mean = 55, sample standard deviation = 10.

c. n = 10, sample mean = 55, sample standard deviation = 5.

d. n = 10, sample mean = 55, sample standard deviation = 3.

15.2 The average chicken pot pie is 8 oz., with 3.25 oz. of chicken. You have taken a sample of 15 chicken pot pies from the assembly line, and you find that the average amount of chicken in the sample is 3.1 oz. Should the assembly line be corrected to produce the correct chicken proportion? The sample standard deviation is 0.3.

15.3 A sample survey finds that 55% of the households in a town were watching the Superbowl. The sample size is 50. Develop a 95% confidence interval for the proportion. Can you be confident that at least 50% of the households in the city were watching the game? With a sample size of 500 and the same data, would your answer to this question be different?

15.4 You are a member of your local city council. You make it a practice to vote for controversial proposals only if an overwhelming majority of your district, which you have defined as 70%, favors them. You face a particularly controversial proposal, and your polling organization (the beginning statistics class at a local university) has done a quick 100 person random sample of your district, finding that 68% of those polled favored the proposal. Could a sample proportion of 0.68 be drawn from a population where the true population proportion is 0.70? Use the one-sample test of proportion calculator menu item to find out. Be sure you define the appropriate hypothesis test. Does it make any difference if you use the normal approximation?

**15.5** You are still on the city council, and you are running for re-election. Your tracking poll is 100 random voters, sampled weekly. Last week's poll shows that you are favored by 52% of the "likely" voters; this week's poll has you at 48%. Is the difference significant? Use the two-sample test of proportion, defining the appropriate hypothesis test.

**15.6** In California, approximately 500,000 valid signatures are necessary to qualify a petition. As the Secretary of State, you have received 600,000 signatures, from which you have taken a random sample of 6,000. You find that 5,075 of the 6,000 are valid. Have the sponsors of the initiative met the requirement? (Use the one-sample test of proportion to check on the proportion of valid signatures (run the test as a binomial test and with the normal approximation); what proportion of the 6,000 have to be valid to ensure that 500,000 valid signatures have been collected?)

**15.7** Consider problem 15.6 with 5,025 valid signatures. 4,975? 4,900? What would your answer be in each of these cases?

**15.8** Develop 95% a confidence interval for the mean using the following sample data: sample mean = 1500, sample standard deviation = 500, sample size = 25. What happens to the confidence interval when the sample size goes from 25 to 100? To 400?

**15.9** Develop a 95% confidence interval around the statistic that a certain city has 200 fatal automobile accidents in a year.

# 16

## Correlation

### The Meaning of Correlation

One aim of research is to discover relationships between variables and to test hypotheses that predict them. In our earlier discussion of scatter plots, for example, we discovered a relationship between average life expectancy and GNP per capita in nations. People tend to live longer, on average, in nations with high GNP than in those with low GNP. Another way to say this is that life expectancy and GNP per capita are **correlated**. It is a **positive** correlation because high life expectancy is associated with high GNP and low life expectancy is associated with low GNP. It is a **strong** correlation because, as we saw in the scatter plot, most of the data points cluster tightly together in a distinct pattern of covariation.

As illustrated by this example, the purpose of bivariate **correlation analysis** is to detect whether a relationship exists between two variables and, if so, to assess its direction (positive or negative) and its strength (strong or weak). SQ's **Correlation** menu offers two kinds of correlation coefficients as tools for such an analysis. Both coefficients range in value from -1.0 (perfect negative correlation) to +1.0 (perfect positive correlation), with a value of zero indicating no correlation. Formulas for computing these coefficients are given in standard statistics texts.

| ■ ■ ■ ■ ■ ■ ■ | ■ **Pearson Correlation Coefficient** |
| --- | --- |
| **IN THIS CHAPTER** | ■ **Spearman Correlation Coefficient** |

The **Pearson coefficient of correlation** (r) indicates the direction and measures the strength of the relationship between two interval-scaled variables. It assumes that the two variables are linearly related—an assumption you should check visually with a scatter plot before computing the coefficient.

The **Spearman coefficient of correlation** ($r_s$) indicates the direction and measures the strength of the relationship between two ordinal (rank-order) variables. It is sometimes also used in correlation analysis of interval-scaled variables because it is resistant to outliers and does not assume the variables are linearly related.

# Pearson Correlation Coefficient

In SQ's *Statistics/Correlation* submenu, the **Pearson (regular)** procedure will compute coefficients of correlation for all possible bivariate relationships in a variable list. For example, if you enter two variables, SQ will compute one coefficient. If you enter ten variables, SQ will report a matrix of (10X9)/2 = 45 coefficients, one for each unique pair of variables. For purposes of inference, the p-value for the sample t-statistic is shown for each coefficient.

### Example (Single Coefficient)

Using the *humandev.dta* file of data on 130 nations, we want to compute a Pearson coefficient of correlation between *life* (average life expectancy in years in 1987) and *literacy* (% adult literacy in 1985). A scatter plot reveals that these two interval-scaled variables are linearly related, so the Pearson coefficient is appropriate to use.

1　From the *File* menu, choose *Open*.

2　Enter the name of the file: humandev.

3　From the *Statistics* menu, choose *Correlation* and then *Pearson (regular)*.

4　**Windows:** In the dialog box, click on *life* and *literacy* as the variable list. Click on OK. Figure 16.1 shows the Pearson Correlation dialog box.

　　**DOS:** In answer to the questions, the variable list should be life and literacy.

The result is shown in Figure 16.2.

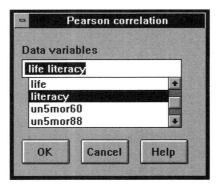

**Figure 16.1** *Dialog Box for Pearson Correlation*

| Stata Results |
| --- |

```
          |     life literacy
----------+------------------
     life |   1.0000
          |
  literacy |   0.8451   1.0000
          |   0.0000
```

**Figure 16.2** *Pearson Correlation Output*

The Pearson coefficient of 0.8451 indicates a strong positive correlation between average life expectancy and adult literacy among nations. The p-value of 0.0000 shown beneath the coefficient could also be expressed as $p < .00005$. It tells us that fewer than one in ten thousand independent samples of this size would produce a Pearson $r$ this large or larger under the null hypothesis of no correlation in the population.

Because the Pearson $r$ is derived from and based on simple linear regression (see Chapter 17), some additional interpretations can be made.

First, in **two-variable** relationships such as this, $r^2 = R^2$, the coefficient of determination in regression. This gives the proportion of variation in Y explained by X. In this example, $r^2 = (0.8451)^2 = .714$, indicating that 71.4% of the variation in life expectancy can be statistically explained by differences in adult literacy.

Second, if the standard deviations of Y and X are known, the Pearson $r$ can be used to compute $b_1$ as an estimate of the slope $\beta_1$ in the equation $Y = \beta_0 + \beta_1 X$. Specifically, $b_1 = r[s_y/s_x]$. Using SQ's *Means and SDs* procedure from the *Summaries* menu, we learn that $s_y$ (*life*) is 10.633 and $s_x$ (*literacy*) is 26.592.

Thus, $b_1$ = .8451 times (10.633 / 26.952) = .338, which implies that average life expectancy increases one-third of a year for every percentage point increase in adult literacy. (You can check this by using SQ's *Statistics/Simple regression* to regress *life* on *literacy*.) This last result shows the added value of including standard deviations of all variables in any report of Pearson correlation coefficients.

## Example (Correlation Matrix)

Using the same data file, we'll now compute Pearson correlation coefficients for all possible bivariate relationships among the variables *life, pop1988, urbpop88, literacy, un5mor88,* and *gnpcap87*.

**1**  From the *File* menu, choose *Open*.

**2**  Enter the name of the file: humandev.

**3**  From the *Statistics* menu, choose *Correlation* and then *Pearson (regular)*.

**4**  **Windows:** In the dialog box, click on *pop1988, urbpop88, life, literacy, un5mor88,* and *gnpcap87* as the variable list. Click on OK.

    **DOS:** In answer to the questions, the variable list should be life, pop1988, urbpop88, literacy, un5mor88, and gnpcap87.

The result is shown in Figure 16.3.

```
 Stata Results

           | pop1988 urbpop88      life literacy un5mor88 gnpcap87
-----------+-------------------------------------------------------
   pop1988 |  1.0000


  urbpop88 | -0.0916   1.0000
           |  0.3000


      life |  0.0740   0.7633   1.0000
           |  0.4028   0.0000


  literacy |  0.0038   0.6704   0.8451   1.0000
           |  0.9659   0.0000   0.0000


  un5mor88 | -0.0547  -0.7160  -0.9575  -0.8455   1.0000
           |  0.5366   0.0000   0.0000   0.0000


  gnpcap87 | -0.0319   0.6522   0.6673   0.5346  -0.6147   1.0000
           |  0.7304   0.0000   0.0000   0.0000   0.0000
```

**Figure 16.3** *Matrix of Pearson Correlation Coefficients*

This matrix reports Pearson correlation coefficients and p-values for each of the 15 distinct relationships among the six variables specified. At the intersection of the column for *life* and the row for *literacy*, for example, is the coefficient of .8451 obtained earlier. Looking at the bottom row of coefficients, we see that *gnpcap87* is positively correlated with *life* (.6673), *urbpop88* (.6522), and *literacy* (.5346). It is negatively correlated with *un5mor88* (-.6147). And it has no evident correlation with *pop1988* (-.0319, very close to zero). Except for this last relationship, all correlations are statistically significant at p < .01.

**Caution:** These conclusions hold only on the assumption of a **linear** relationship between the variables analyzed. That assumption is reasonable in the case of *life* and *literacy*. We checked that with a scatter plot. If we run a scatter plot of the relationship between *life* and *gnpcap87*, however, we can see that the linearity assumption is false. The relationship between these two variables is distinctly nonlinear. The reported Pearson *r* of .6673 for the relationship between *life* and *gnpcap87* is therefore suspect. In fact, it dramatically **under**estimates the real strength of the (nonlinear) relationship between those two variables. (The Spearman coefficient, because it does not **assume** a linear relationship, offers a more trustworthy measure of the true correlation in this case.)

**Advice:** Run a scatter plot to check that a bivariate relationship really is at least approximately linear *before* computing a Pearson correlation coefficient. If you are "matrix minded" in your approach to analysis, first inspect a scatter plot matrix of the variables to be analyzed (see Chapter 10). Then run the correlation matrix, keeping in mind the picture that lies behind each coefficient. **Related advice:** Hypothesis tests should be conducted one at a time and in the context of linear regression, which is the basis for computing coefficients. Avoid searching through a matrix for just those coefficients with p-values less than .05, discarding the rest. If you must conduct whole batches of such tests in matrix form, use the Bonferroni or Sidak corrections (see Chapter 21 on ANOVA).

# Spearman Correlation Coefficient

The Spearman coefficient of correlation $(r_s)$ is a nonparametric (distribution-free) measure of association between two rank-ordered variables. Unlike the Pearson coefficient, the Spearman coefficient does not assume a linear relationship and is resistant to the distorting effects of outliers. Because observations on any variable measured at a higher (interval or ratio) level can be sorted and ranked, the Spearman coefficient could and probably should be used more frequently than it is.

## Example (Rank-Ordered Data)

The *cities2.dta* file contains information reported by *Fortune* magazine on business conditions in 60 large U.S. cities (Labich, 1993). As judged by business experts, the variable *innov* ranks each city on its "presence of innovative firms," and the variable *probiz* ranks each city on its "pro-business attitude." For both variables, the highest-ranking city is scored 1 and the lowest ranking city is scored 60. We hypothesize that these two variables are positively correlated. It is appropriate to use the Spearman rank-order correlation coefficient to measure the strength of association between these two ordinal variables.

**1** From the *File* menu, choose *Open*.

**2** Enter the name of the file: `cities2`.

**3** From the *Statistics* menu, choose *Correlation* and then *Spearman (rank)*.

**4** **Windows:** In the dialog box, click on *innov* to select variable 1 and *probiz* to select variable 2. Click on OK. Figure 16.4 shows the Spearman rank correlation dialog box.

   **DOS:** In answer to the questions, the variable 1 should be `innov` and variable 2 should be `probiz`.

The result is shown in Figure 16.5.

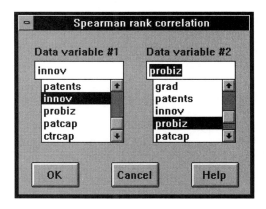

**Figure 16.4** *Dialog Box for Spearman Correlation Coefficient*

```
Stata Results

spearman innov probiz

 Number of obs =       60
Spearman's rho =      0.2181

Test of Ho: innov and probiz independent
     Pr > |t| =      0.0942
```

**Figure 16.5** *Spearman Correlation Output*

The Spearman coefficient of .2181 confirms the hypothesized positive correlation between pro-business attitude and presence of innovative firms in large U.S. cities. The correlation is weak, however, and the p-value is a rather large .0942. If this were a sample, one would not reject the null hypothesis using the conventional $\alpha$ of .05.

## Example (Interval-Scaled Data)

Earlier we obtained a Pearson correlation coefficient of .6673 for the relationship between life expectancy and GNP per capita in nations. We also argued that this coefficient underestimated the real strength of the correlation between these two variables on the false assumption that they are linearly related. The Spearman coefficient, which requires no such assumption, should give a better result. Let's see if this is true.

**1**   From the *File* menu, choose *Open*.

**2**   Enter the name of the file: humandev.

**3**   From the *Statistics* menu, choose *Correlation* and then *Spearman (rank)*.

**4**   **Windows:** In the dialog box, click on *life* to select variable 1 and *gnpcap87* to select variable 2. Click on OK.

   **DOS:** In answer to the questions, the variable 1 should be life and variable 2 should be gnpcap87.

The result is shown in Figure 16.6.

```
Stata Results

. spearman life gnpcap87

 Number of obs =       119
Spearman's rho =       0.8596

Test of Ho: life and gnpcap87 independent
     Pr > |t| =       0.0000
```

**Figure 16.6** *Spearman Correlation Output-Interval Level Data*

As predicted, the Spearman coefficient of .8596 is much larger than the Pearson, reflecting the strong correlation that is obvious in a scatter plot of these two variables.

> ## ■ TIP 18 ■ ■ ■ ■ ■ ■ ■ ■ ■ ■ ■ ■ ■ ■ ■
>
> ### WARNING—SQ Sorts the Data When It Runs the Spearman Correlation
>
> To compute the Spearman coefficient, StataQuest sorts the data set from low to high on the second variable entered and does not restore the original order. If it is important to you, be sure to have some way (for example, unique ID numbers) to resort your data to its original order before running the Spearman procedure. If that's not possible, use the procedure, but do not save and replace.

## ■ EXERCISES

**16.1**   Open the *highway3.dta* file.

    **a.**   Produce a scatter matrix showing relationships among the variables *rate-fata*, *exceed55*, *exceed65*, and *popdense*. Interpret your results. Which relationships seem appropriate for computing Pearson coefficients, and which not? Why?

    **b.**   Keeping in mind the results of (a), produce a matrix of Pearson coefficients for these four variables. Interpret the results.

    **c.**   If you have reservations about using the Pearson for any of these relationships, what are they? What other approaches to correlation analysis might you pursue?

**16.2**   Use *cities2.dta* to make the following analyses.

    **a.**   Compute Spearman coefficients of correlation between *innov* and the following: *ctrcap*, *patcap*, and *ba*. Conclusions?

    **b.**   Browse the spreadsheet or use the *List data* procedure to identify cities that are ranked low on innovation but high on one or more of the other variables or vice versa. What clues do these listings provide to guide additional research?

**16.3**   (More advanced) Often a nonlinear bivariate relationship can be "linearized" by transforming one or both of the variables to a different scale. That would allow procedures such as simple regression and Pearson correlation to be used. Open the *humandev.dta* file.

    **a.**   Run a scatter plot of *life* vs *gnpcap87*. Note the relationship.

b.  Obtain the Pearson coefficient of correlation between these two variables.

c.  Run a scatter plot of *life* vs. *loggnp* (base-10 logarithm of GNP per capita). Does this relationship look approximately linear?

d.  Obtain the Pearson coefficient for these two variables. Compare with the Pearson coefficient obtained in (b) and with the Spearman coefficient of .8596 reported in the text. Conclusions?

**16.4**  (More advanced) After clearing memory, generate 20 normally distributed variables X1, X2, ...X20 with mean=0 and standard deviation=1 for 200 observations.

a.  Produce a Pearson correlation matrix for these 20 variables.

b.  Count the number of coefficients that are "statistically significant" at $p < .05$.

c.  Divide the count obtained in (b) by 190, the number of unique coefficients in the matrix. Is the fraction close to 5%? What do you learn from this?

# 17

■ ■ ■ ■ ■ ■ ■

# *Simple Regression and Post-Regression Diagnostics*

## Modeling Two-Variable Linear Relationships

StataQuest's **simple regression** computes the ordinary least-squares (OLS) regression of a dependent Y variable on a single independent X variable. It fits the model Y = b*X + c + error by estimating the values of b (the slope) and c (the constant) that best fit your data on the assumption that Y and X are linearly related. The result is called the regression equation. An example is Y = 4.43X + 2370. This equation, derived from an analysis of hospital survey data (*lowbw.dta*, on disk), summarizes the linear relationship between infant birth weight in grams (Y) and mother's weight in pounds (X) observed in a sample of 189 births. The estimated slope (b) is 4.43. The estimated constant (c) is 2370 grams. In addition to estimating the slope and constant, SQ's simple regression computes goodness of fit measures, standard errors, confidence intervals, and other statistics used for inference (for example, testing hypotheses, estimating population parameters). In the birth weight study, for example, the standard error of the slope is 1.71 and the t-statistic is 2.71, indicating statistical significance at p < .01.

**235**

---

- **Simple regression**
- **Post-regression diagnostics**

---

SQ's optional **post-regression diagnostics** include a plot of the model fit (Y versus X with regression line); a plot of residuals versus predicted values (Y-residuals versus Y-hats); a plot of residuals versus another X; and a normal quantile plot of residuals. You can also save the residuals and predicted values in your data file.

## Simple Regression

SQ's **simple regression** is limited to one independent variable. It assumes that Y is measured on at least an interval scale and that Y and X are linearly related. It assumes that the data analyzed are free of severe outliers. Many more assumptions are required if you want to generalize your results from a sample to a population. One such assumption, for example, is that the error term (representing the net effects on Y of all variables excluded from the model plus the effects of random disturbances) is normally distributed around a mean of zero. These kinds of technical assumptions and the mathematical formulas underlying them are beyond the scope of this book.

Simple regression can be applied to many different kinds of data and analytical situations. You can use it with a categorical independent variable and thus as an alternative to **ANOVA** (Chapter 20). With appropriate measurement transformations (for example, log transforms), you can often use it to model and analyze nonlinear relationships. **Multiple regression** (Chapter 18) extends simple regression to include two or more independent variables.

Simple regression does not perform well in the presence of outliers, however, and in that situation you might consider using **robust regression** (Chapter 18). Nor should this method be used with a categorical dependent variable. In the case of a dichotomous categorical dependent variable with one independent variable, consider using **logistic regression** (Chapter 19).

### Running a Simple Regression: Example

SQ makes it very easy to do a simple regression. Open a data set, specify a Y and an X, and SQ will give you the results in seconds. The hard part is knowing which Y and which X to put in (theory) and making sense of the statistics and graphs that come out (interpretation). We'll leave the theory to you. Our task here is to show how to run a simple regression and interpret the output.

To illustrate SQ's simple regression procedure, we will analyze the relationship between average life expectancy and urbanization in the world's nations. The data for 130 nations are contained in the *humandev.dta* file (on disk). The Y variable of interest is *life* (average life expectancy in years at birth in 1987). The X variable is *urbpop88* (percentage of population living in urban areas in 1988).

Before using SQ to regress *life* on *urbpop88*, we should do three things: (1) Conduct a complete univariate analysis of both *life* and *urbpop88* to check for outliers and to be sure the measurement and distributional assumptions of regression are satisfied. (2) Inspect a scatter plot of *life* versus *urbpop88* to confirm visually that the relationship between them is approximately linear. (3) Open a log file to save all nongraphical regression output on disk. Let's assume these steps have been taken.

**1**  From the *File* menu, choose *Open*.

**2**  Enter the name of the file: humandev.

**3**  From the *Statistics* menu, choose *Simple regression*.

**4**  **Windows:** Click on the variable named *life* as the dependent variable and on the variable named *urbpop88* as the independent variable.

   **DOS:** Enter the variable named life as the dependent variable and the variable named urbpop88 as the independent variable.

The results appear in Figure 17.1.

```
Stata Results

  Source |       SS       df       MS           Number of obs =      130
---------+------------------------------         F( 1,   128) =   178.64
   Model | 8496.46113       1  8496.46113         Prob > F      =   0.0000
Residual | 6088.0081      128  47.5625633         R-squared     =   0.5826
---------+------------------------------         Adj R-squared =   0.5793
   Total | 14584.4692     129  113.057901         Root MSE      =   6.8966
----------------------------------------------------------------------------

    life |    Coef.   Std. Err.      t     P>|t|     [95% Conf. Interval]
---------+------------------------------------------------------------------
urbpop88 | .3275431  .0245065    13.366    0.000     .2790527    .3760335
   _cons | 46.61286  1.348095    34.577    0.000     43.94542    49.2803
```

**Figure 17.1** *Output of Simple Regression of Average Life Expectancy in Years at Birth in 1987 on Percentage Urban Population in 1988: 130 Nations*

## Interpreting the Output

The regression output displayed in Figure 17.1 consists of **three subtables**, one at the upper left, one at the upper right, and one at the bottom. Starting with the upper-left subtable, read the annotations in notes A through D that follow. Then continue with the upper right and bottom subtables that follow.

### Upper-Left Subtable

```
  Source |      SS        df       MS             ← A
---------+----------------------------------
   Model |  8496.46113      1   8496.46113        ← B
Residual |   6088.0081    128   47.5625633        ← C
---------+----------------------------------
   Total |  14584.4692    129   113.057901        ← D
```

A. Headings: **Source** refers to the partitioning of the total sum of squares (TSS) into its component parts, the model (explained sum of squares or ESS) and the residual (residual sum of squares or RSS). Note that Total = Model + Residual, or TSS = ESS + RSS. The **SS** denotes the sum of squares, **df** stands for degrees of freedom, and **MS** labels the mean sum of squares (sum of squares divided by degrees of freedom).

B. **Model** sum of squares (ESS) is 8496.46113 with $K - 1 = 1$ degree of freedom. K is the number of parameters to be estimated, which is two in the case of simple regression: the slope and intercept. The MS is 8496.46113, which is ESS/df or 8496.46113 divided by 1. This is the numerator in the F-statistic.

C. **Residual** sum of squares (RSS) is 6088.0081 with $N - K = 128$ degrees of freedom. (N is the number of observations = 130, K the number of parameters to be estimated = 2.) MS is 47.5625633, which is RSS/df or 6088.0081 divided by 128. This is the denominator in the F-statistic.

D. **Total** sum of squares (TSS) is 14584.4692 with $N - 1 = 129$ degrees of freedom. MS is 113.057901, the sample variance of Y. This is the variation in Y we seek to explain with X.

### Upper-Right Subtable

```
Number of obs =       130      ← E
F(  1,    128) =   178.64      ← F
Prob > F      =   0.0000      ← G
R-square      =   0.5826      ← H
Adj R-square  =   0.5793      ← I
Root MSE      =   6.8966      ← J
```

E. The total **number of observations** is 130.

F. The **F-statistic** is 178.64 at 1 and 128 degrees of freedom. This equals the Model MS divided by the Residual MS, or 8496.46113 divided by 47.5625633. In the bivariate case only it also equals the t-statistic squared, or 13.366 times 13.366. The F-statistic is used to test the null hypothesis that all slopes in the population are equal to zero. (The F-statistic is used mainly to evaluate the results of multiple regression.) The F-statistic is evaluated against critical values for the F-distribution at K − 1 (numerator) degrees of freedom and N − K (denominator) degrees of freedom.

G.  Stated informally, **Prob > F = 0.0000** means that there is less than one chance in ten thousand that an F-statistic as large or larger than 178.64 could have been obtained for a sample of this size from a population in which the regression slope is zero.

H.  The **R-square** ($R^2$) of .5826 is the proportion of variation in Y attributed to X. It is the ratio of the model sum of squares (ESS) to the total sum of squares (TSS) or 8496.46113 divided by 14584.4692. Called the coefficient of determination, it is often used as a measure of a regression model's goodness-of-fit with the sample data. In simple (bivariate) regression, the square root of R-square is the Pearson correlation coefficient. Thus Pearson's r = .763, the square root of .5826. In the bivariate case only, the Pearson's r of .763 is also equal to the standardized regression slope (sometimes called the beta weight) obtained from regressing z-scores for Y on z-scores for X. This tells us that a one standard deviation unit increase in X is associated with a .763 standard deviation unit increase in Y.

I.  The **Adj R-square** of .5793 is the adjusted R-square. It is the statistic of choice in reporting goodness-of-fit for models with two or more independent variables. The R-square can always be increased simply by adding more X variables to the model. The adjusted R-square is R-square adjusted for the ratio of the number of observations to the number of coefficients. Its use encourages parsimony by discounting small improvements in explained variance obtained at the price of adding new variables and complexity. Unlike the R-square, the adjusted R-square can actually decrease in value as more variables are added and can even be negative. Click on Help in the dialog box for *Multiple regression* to see the correction formula.

J.  The **Root MSE** of 6.8966 is the square root of the mean squared residuals (errors). Thus the square root of 47.562563 is 6.8966. It is also called the sample standard deviation of the residuals. It is used extensively in regression diagnostics and the analysis of residuals.

### *Bottom Subtable*

```
-------------------------------------------------------------------
    life |     Coef.   Std. Err.      t     P>|t|    [95% Conf. Interval]  ← K
---------+---------------------------------------------------------------
urbpop88 | .3275431    .0245065   13.366    0.000     .2790527   .3760335  ← L
   _cons | 46.61286    1.348095   34.577    0.000     43.94542    49.2803  ← M
```

K. Headings: **life** is the name of the dependent Y variable. **Coeff** labels the column of regression coefficients that estimate the slope and intercept. **Std. Err.** labels the column of estimated standard errors for the slope and intercept. **t** labels the column of t-statistics for the slope and intercept. **P>|t|** labels the column of p-values for the t-statistics. **[95% Conf. Interval]** labels the column of 95% confidence intervals for slope and intercept coefficients.

L. **urbpop88** is the name of the independent X variable. The **slope coefficient** is .3275431, the estimate of b in the model Y = b*X + c + error. Under a causal interpretation, this slope estimate indicates that average life expectancy increases about one-third of a year for every one percentage point increase in the urban population. The **standard error of the slope** is .0245065. The **t-statistic** is 13.366, which is the ratio of the slope to its standard error. The **probability** is less than one chance in a thousand (0.000) of getting a t-statistic with an absolute value this large or larger for a sample of this size under the null hypothesis of a zero slope in the population. The **confidence interval** indicates that we can be 95% sure that the true value of the slope in the population lies between .2790527 and .3760335. Note that zero is not included in this interval.

M. **_cons** identifies the **constant** (also known as the **intercept**) in the regression equation. The constant is 46.61286, an estimate of the c in the model Y = b*X + c + error. The **standard error of the constant** is 1.348095. The constant is rarely of much theoretical interest in research and often has no substantive meaning. In this analysis, it does mean something. Specifically, the estimated average life expectancy in totally rural populations (in which *urbpop88* = 0) is 46.61 years. The remaining statistics can be interpreted similarly to those in (L).

***Summary***

The model fit obtained from simple regression is Y = .3275431*X + 46.61286. Rounding off, this regression equation tells us that the expected value (mean) of average life expectancy conditional on percentage urban is 46.61 years plus .328 times the percentage urban. This relationship is statistically significant at $p < .001$. The level of urbanization statistically explains about 58% of the variation in average life expectancy among nations.

## Using the Regression Equation to Compute Predicted Values and Residuals

We can use the regression equation Y = .3275431*X + 46.61286 to compute the predicted average life expectancy of each nation's population given its percentage urban. In StataQuest, these **predicted values of Y** from the regression equation are called **Y-hats**.

To illustrate, we substitute 21 for X in the regression equation, and we get .3275431(21) + 46.61286 = 53.49 years (rounded to two decimals). The result of 53.49 years is the predicted **mean** of average life expectancy for populations that are 21% urban. Sri Lanka's percentage urban in 1988 was 21%, so we'll use 53.49 years as our best guess (based on the regression analysis) of what the average life expectancy of Sri Lanka's population "should" be given its percentage urban. The Y-hat for Sri Lanka, therefore, is 53.49 years.

The difference between the actual value and the predicted value (Y minus Y-hat) for each observation is called the **residual**. The residual can be either positive or negative.

For example, the actual average life expectancy of Sri Lanka's population (Y) is 71 years. The predicted value (Y-hat) is 53.49 years. Therefore, the residual for Sri Lanka is 71 minus 53.49 = +17.51 years. The average life expectancy of Sri Lanka's population is much larger than it "should" be (based on the regression equation) given its low level of urbanization.

Reported here are the three nations with the lowest negative residuals and the three with the highest positive residuals.

| name | urbpop88 | ACTUAL life | PREDICTED life_hat | RESIDUAL life_res |
|---|---|---|---|---|
| Central Afr Rep | 45 | 46 | 61.3523 | -15.3523 |
| Sierra Leone | 31 | 42 | 56.7667 | -14.7667 |
| Benin | 40 | 47 | 59.71458 | -12.71459 |
| China | 21 | 70 | 53.49127 | 16.50873 |
| Portugal | 32 | 74 | 57.09424 | 16.90576 |
| Sri Lanka | 21 | 71 | 53.49127 | 17.50873 |

These statistics—the Y's, the Y-hats, and the residuals—all play a key role in post-regression diagnostics.

## Post-Regression Diagnostics

After running a simple regression of Y on X and studying the results, you may want to extend your analysis in several ways.

A natural first step would be to **plot the model fit** by making a scatter plot of Y versus X with the regression line. You could then see whether the regression line of predicted values from the equation actually fits the overall pattern of the data revealed by the plot.

A useful second step would be to **plot the residuals against the predicted values**. With this diagnostic graph you can check for certain visual symptoms of "unhealthy" residuals—for example, heteroscedasticity (nonconstant variance), the presence of severe outliers, and nonrandom structure of any kind.

A third step would be to **run a normal quantile plot of the residuals** to see if they are indeed normally distributed as required by regression for purposes of inference. If they are not, the computed standard errors, confidence intervals, and t-tests all become untrustworthy for generalizing sample results to a population.

A fourth step you might want to take, especially if you're in an exploratory mood, is to **plot the residuals against a new X variable**. The residuals, by definition, measure the variation in Y unexplained by the first X in your model. So if this plot shows a relationship between the residuals and your new X, you might consider including it as a second independent variable in a new round of analysis using SQ's *Multiple regression* procedure.

SQ makes it easy for you to take each of these steps simply by clicking on a button in the *Post-regression diagnostics* menu.

## Running the Post-Regression Diagnostics: Example

We'll now extend our regression analysis of the relationship between average life expectancy and urbanization in nations by moving it into the post-regression diagnostics stage. First, let's repeat the regression command, this time asking SQ to show the diagnostics menu.

**1**  From the *File* menu, choose *Open*.

**2**  Enter the name of the file: `humandev`.

**3**  From the *Statistics* menu, choose *Simple regression*.

**4**  **Windows:** Click on the variable named *life* as the dependent variable and on the variable named *urbpop88* as the independent variable. Click on the *Show diagnostics* menu option box.

   **DOS:** Enter the variable named `life` as the dependent variable and the variable named `urbpop88` as the independent variable. Request the diagnostics menu.

The regression results were reported earlier in Figure 17.1. The requested post-regression diagnostics menu is shown in Figure 17.2.

With the diagnostics menu available, we'll now ask for each of the diagnostic plots in succession and interpret the graphical output. **To run a diagnostic, click on the radio button to the left of the menu item and then click on *Run*.** **DOS**: select the desired diagnostic from the menu, press the ENTER key, and then answer any prompts by following the guidelines shown for the Windows version.

### Plot Fitted Model

When you select this diagnostic plot, a **second pop-up menu** will ask if you want the plot to include a *confidence interval* and/or a *prediction interval*. If

## ■ TIP 19 ■ ■ ■ ■ ■ ■ ■ ■ ■ ■ ■ ■ ■ ■ ■ ■

### Arranging Your Windows to Display Diagnostic Plots

When requested, the *Post-regression diagnostics* menu shown in Figure 17.2 will be brought to the front and will cover part of the regression output displayed in the Results window. Later, as you begin running the various diagnostic plots, the diagnostics menu will always be brought to the front, partially eclipsing the plot displayed in the Graph window. You can always click on the Graph button to bring the Graph window to the front, and then click on the Dialog button to restore the diagnostics menu. A better solution, however, is to **push the Graph Window out of the way**. With the mouse, move your cursor to the left border of the Graph Window. When the cursor turns into a double-arrow, click and drag the border toward the right just far enough to reveal the right edge of the diagnostics menu. Then release the mouse key. The Graph window will be a bit smaller than before, but now you can click on the various diagnostic plots and see instant results without interruption. (You can also make a little more room for the plots by clicking and dragging on the title bar of the diagnostic menu to move it to the left.) You will have to undo all of this, of course, after you're through with the diagnostics.

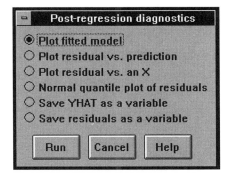

**Figure 17.2** *Optional Dialog Box for Post-Regression Diagnostics Following Simple Regression*

you select either or both, you can accept the displayed default of 95% or type the desired significance level (for example, 90 or 99). Selection of the *confidence interval* option will draw confidence bands around the regression line for predicting the **mean value of Y** conditional on X at the specified significance level. Selection of the *prediction interval* option will draw confidence bands around the regression line for predicting **individual Y values** conditional on X at the specified significance level. Both intervals can be displayed at the same time on the plot. The prediction interval will always be wider than the confidence interval around the regression line. If you choose neither, only the regression line will be drawn.

Let's ask for a diagnostic plot that shows the fitted model with a 95% confidence interval.

In the *Post-regression diagnostics* menu that appears following the simple regression of *life* on *urbpop88*:

**1** Click on *Plot fitted model*.

**2** In the pop-up menu that appears, click on the *Include confidence interval* option box.

**3** Accept the significance level of 95%. If some other number is in the box, type 95.

**4** Click on *Run*.

The results appear in Figure 17.3.

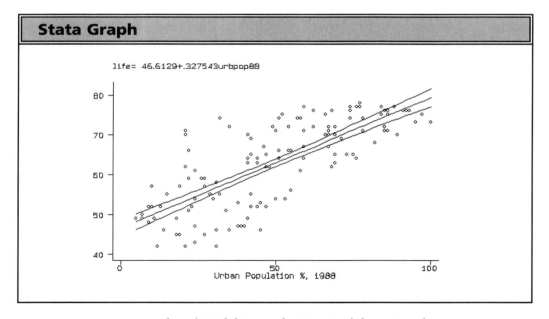

**Figure 17.3** *Plot of Model Fit with 95% Confidence Bands*

Except for the addition of the confidence lines, the graph in Figure 17.3 is identical to that of the scatterplot with regression line available in the *Graphs—Scatterplots* menu. The regression line is drawn on a scatter plot of *life* versus *urbpop88*, and the regression equation is printed at the top of the graph. The 95% confidence lines drawn on either side of the regression line show the 95% confidence interval for the predicted mean average life expectancy at each level of percentage urban. Notice that the confidence interval is narrowest near the mean of *urbpop88* (49%) and widest at the extremes. Overall, the plot shows that the linear regression model fits the data fairly well.

### Plot Residual versus Prediction

When you run a simple regression and ask for diagnostics, SQ uses the regression equation to generate two **temporary** variables named *_Yhat* (predicted mean values of Y conditional on X) and *_Resid* (the residuals). (Note: The content of these variables changes from regression to regression. If you want to save a particular *_Yhat* or *_Resid* permanently in your data set, use one or both of the last two options in the *Post-regression diagnostics* menu. See the following.) The *Plot residual vs. prediction option* in the diagnostics menu will plot *_Resid* versus *_Yhat* based on the last-run regression.

In the *Post-regression diagnostics* menu that appears following the simple regression of *life* on *urbpop88*:

**1**  Click on *Plot residual vs. prediction.*

**2**  Click on *Run.*

The results appear in Figure 17.4.

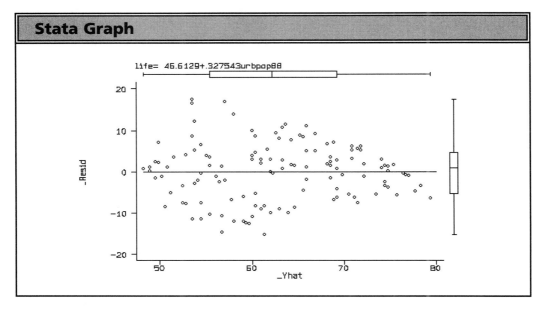

**Figure 17.4** *Plot of Residuals (_Resid) Versus Predicted Values (_Yhat) from the Regression Equation.*

First, let's focus on rather ingenious design of this plot. The residuals in Figure 17.4 are plotted on the Y axis and the predicted values on the X axis. A box plot of the residuals is displayed on the right side of the graph and a box plot of the predicted values is displayed at the top. The box plots indicate that both the residuals and the predicted values are symmetrically distributed with no outliers. The regression equation is printed at the very top of the graph. The horizontal line drawn from the zero point on the Y axis effectively flattens the regression line and makes it easy to distinguish the negative residuals (below the line) from the positive residuals (above the line). This flattening effect also reduces the likelihood of perceptual illusions, such as the tendency to underestimate the magnitude of residuals for points that are close to a steeply sloped line.

Are the residuals displayed in Figure 17.4 "healthy" or "unhealthy" by conventional standards? Healthy residual plots should look patternless. Allowing for the quirks of true randomness, there should be about as many negative residuals as positive residuals. The residuals should be distributed with approximately the same variance across the range of predicted values. No outliers or distinct clusters of cases should stand out apart from the rest. By most of these criteria, the residuals plotted in Figure 17.4 look fairly healthy.

There is one exception. Notice the clear pattern of diminishing variance in the residuals at higher levels of predicted values on the X-axis. This violates the assumption of constant variance across the range of X—a condition statisticians call heteroscedasticity. Standard errors are based on a single standard deviation of the residuals—an average over the whole set. As shown in Figure 17.4, the standard error of the residuals will be too small in the lower range of X where the variance is large and too large in the higher range of X where the variance is small. For purposes of inference, the standard error of the slope, statistical tests, and confidence intervals may be somewhat inaccurate as a result. (Intermediate and advanced texts prescribe weighting corrections and robust regression methods to deal with heteroscedasticity. See, for example, Hamilton, 1992.)

### Normal Quantile Plot of Residuals

One way to assess whether the errors in the model $Y = b * X + c + error$ are distributed approximately normally, as required for inference, is to study a normal quantile plot of the residuals from the regression equation. In the normal quantile plot (see Chapter 6), if the residuals for each observation lie fairly close to the line of equality, we can be more confident that the errors are normally distributed.

In the *Post-regression diagnostics* menu that appears following the simple regression of *life* on *urbpop88*:

**1** Click on *Normal quantile plot of residuals.*

**2** Click on *Run.*

The results appear in Figure 17.5.

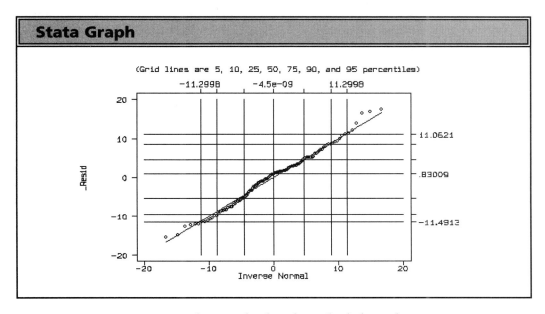

**Figure 17.5** *Normal Quantile Plot of Residuals from the Regression Equation*

The normal quantile plot displayed in Figure 17.5 provides clear visual evidence that the residuals are normally distributed. (Also see Exercises 17.1 and 17.2.)

### Plot Residual versus An X

Based on our analysis thus far, the evidence suggests that average life expectancy and urbanization are positively and linearly related. Theory suggests that other factors besides urbanization contribute to an increase in average life expectancy. One such factor is a nation's level of economic development, as measured by the base-10 logarithm of its Gross National Product per capita. In the *humandev.dta* file, this variable is named *loggnp*.

If economic development has an independent effect on average life expectancy, we would expect to find a relationship between *loggnp* and the temporary variable *_Resid*, which measures the residual variation in *life* that is **unexplained** by *urbpop88* under the linear model. We can conduct a rough visual test of this hypothesis by selecting the *Plot residual vs. an X* option in the *Post-regression diagnostics* menu and entering *loggnp* as the new X.

In the *Post-regression diagnostics* menu that appears following the simple regression of *life* on *urbpop88*:

**1**   Click on *Plot residual vs. an X.*

**2**   In the pop-up menu that appears, click on the independent variable *loggnp* as your X.

**3**   Click on Run.

The results appear in Figure 17.6.

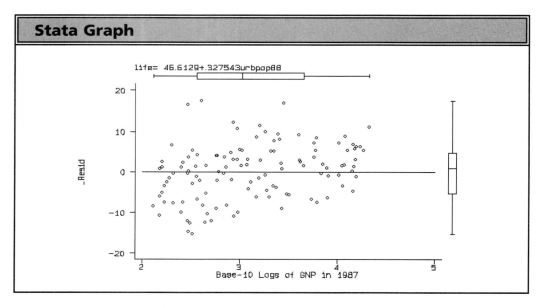

**Figure 17.6** *Plot of Residuals (_Resid) versus Base-10 Logarithms of GNP per Capita in 1987*

The plot of *_Resid* versus *loggnp* in Figure 17.6 shows a positive and roughly linear relationship between these two variables. A reasonable next step in the analysis would be to switch to SQ's *Multiple regression* procedure and regress *life* on **two** independent variables, *urbpop88* and *loggnp*. (See Chapter 18.)

### Saving Predicted Values and Residuals as Variables in Your Data Set

As illustrated by these examples, the predicted values and residuals from a simple regression are essential for detailed post-regression diagnostic analysis. You might want to save these temporary variables permanently in your data file for later use. The *Save YHAT as a variable* and *Save residuals as a variable* options in the *Post-regression diagnostics* menu make it easy to do

┌─────────────────────────────────────────────────────────────────────┐

■ **TIP 20** ■ ■ ■ ■ ■ ■ ■ ■ ■ ■ ■ ■ ■ ■ ■ ■

## AV Plots as an Alternative to Plot Residual versus an X

The *Plot residual vs. an X* option provides a useful graphical tool for exploratory analysis and first-stage model building. Outside the menus, however, SQ's **AV plots** (added-variable plots, also known as partial-regression leverage plots) offer a more powerful and reliable graphical tool for this kind of exploratory model-building. AV plots and other post-regression diagnostic tools, including an entire battery of influence statistics, are available following use of SQ's *fit* command. The *fit* command and AV plots will be discussed briefly in Chapter 18. For more detailed information, click on Help in the Multiple regression dialog box.

└─────────────────────────────────────────────────────────────────────┘

so. Subject to SQ's limits on the number of variables and observations that can be stored in one data file, you can save the predicted values and residuals that interest you, each under a different name.

**To save the predicted values from the regression** as a permanent variable, click on the *Save YHAT as a variable* option. In the pop-up menu that appears, type a name for the new variable (maximum 8 characters) or accept the default name shown in the box. One suggestion is to include the term "hat" in the variable name to identify it as a Y-hat from a regression. If you plan to run more regressions on the same dependent variable, you might add a number to keep track. For example, you might save three different predicted values for the variable *life* as *lifehat1*, *lifehat2*, and *lifehat3*. Be sure to save and replace the file with these new variables before exiting SQ.

**To save the residuals from the regression** as a permanent variable, click on the *Save residuals as a variable* option. In the pop-up menu that appears, type a name for the new variable (maximum 8 characters) or accept the default name shown in the box. One suggestion is to include the term "res" in the variable name to identify it as a residual from a regression. If you plan to run more regressions on the same dependent variable, you might add a number to keep track. For example, you might save three different residuals for the variable *life* as *liferes1*, *liferes2*, and *liferes3*. Be sure to save and replace the file with these new variables before exiting SQ.

Once saved as permanent variables in your data file, the Y-hats and residuals can be analyzed and plotted using other SQ tools. For example, rather than relying only on the normal quantile plot to check whether the residuals are normally distributed, you can run a histogram of the residuals with a normal curve overlay. As another example, you might (1) save the Y-hat from regressing *life* on *urbpop88* using simple OLS regression as *yhatOLS*, and (2) save the Y-hat from regressing *life* on *urbpop88* using robust regression (see Chapter 18) as *yhatROB*. You could then plot **both** model fits on the **same** scatter plot of *life* versus *urbpop88* by typing the following command in SQ's Command window: `graph life yhatOLS yhatROB urbpop88, c(.11) s(Oii) sort`. (Note: Those are small "L"s in the c( ) connect option.) The result would be a good diagnostic plot for detecting the effects of outliers, if any, on the OLS model fits.

## ▨ EXERCISES

**17.1**   Open the *humandev.dta* file (on disk).

   a.   Run a simple regression of the variable named *life* on the variable named *gnpcap87*. Be sure to ask for the diagnostics menu.

   b.   In the diagnostics menu, run the following: (1) Plot of model fit; (2) Plot of residual versus prediction; (3) Normal quantile plot of residuals.

   c.   Save the predicted values from the regression as a variable named *yhat1*.

   d.   Save the residuals from the regression as a variable named *yres1*. Questions: What is the regression equation? Slope? Standard error of the slope? Confidence interval for the slope? Adjusted R-square? What does the diagnostic plot of model fit reveal? Would you say the relationship observed is linear? What does the diagnostic plot of residuals reveal? Would you say the residuals are "healthy"? Based on the normal quantile plot, would you say that the residuals are normally distributed?

**17.2**   Open the *humandev.dta* file (on disk).

   a.   Run a simple regression of the variable named *life* on the variable named *loggnp*. Be sure to ask for the diagnostics menu.

   b.   In the diagnostics menu, run the following: (1) Plot of model fit; (2) Plot of residual versus prediction; (3) Normal quantile plot of residuals.

   c.   Save the predicted values from the regression as a variable named *yhat2*.

   d.   Save the residuals from the regression as a variable named *yres2*. Questions: What is the regression equation? Slope? Standard error of the slope? Confidence interval for the slope? Adjusted R-square? What does the diagnostic plot of model fit reveal? Would you say the relationship observed is linear? What does the diagnostic plot of residuals reveal? Would you say the residuals are "healthy"? Based on the normal quantile plot, would you say that the residuals are normally distributed?

**17.3**   (Note: This exercise assumes that you have done Exercises 17.1 and 17.2 and that you have saved the variables *yhat1*, *yhat2*, *yres1*, and *yres2* in the *humandev.dta* file.) Open the *humandev.dta* file (on disk).

a.   Use the *Graphs–One variable–Histogram–Continuous variable* command (see Chapter 6) to run a histogram of *yres1* with 12 bins and normal curve overlay. Do the same for *yres2*. Compare the results.

b.   Use the *Graphs–Scatterplots–Plot Y vs. X naming points* command to plot *yres2* versus *yhat2* with the variable named *name* as the labeling variable. Do you notice anything unusual about China and Sri Lanka? How about Oman and Gabon?

c.   In the Command Window, type `graph life yhat1 yhat2 gnpcap87`. Which of the two predictions (both conditional on GNP per capita) fit the data better, *yhat1* or *yhat2*?

**17.4**   Open the *anscombe.dta* file (on disk). This file contains the famous "Anscombe Quartet" of artificial data sets developed years ago by F. Anscombe to dramatize the dangers of doing regression analysis without looking at your data. Perversely, this exercise will have greatest instructional impact if you follow the bad practice of leaping (into regression) before you look.

a.   In succession, (1) regress *y1* on *x1*, (2) regress *y2* on *x2*, (3) regress *y3* on *x3*, and (4) regress *y4* on *x4*. Do **not** request the diagnostics menu. After each regression, study the output carefully. By the third or fourth regression you will notice a certain pattern in the results.

b.   Repeat all of the regressions in (a), only this time **do** ask for the diagnostics menu and plot the fitted model after each run. What conclusions do you draw from this exercise?

**17.5**   (Advanced) Open the *planets.dta* file (on disk). This file contains various data on the nine planets in our solar system.

a.   Regress *revol* (planet's revolutionary period in earth days) on *distsun* (average distance from the sun in millions of miles). Study the output and run the appropriate diagnostics. Would you accept the linear model as a good summary of the observed relationship? Why or why not?

b.   Now regress *revol2* (*revol* squared) on *dist3* (*distsun* cubed). Study the output and run the diagnostics. What evidence do you see in your results to support the claim made by Johannes Kepler in 1619 that the "ratio of the cube of the semimajor axis to the square of the period is the same for all the planets including the earth"?

# 18

# *Multiple Regression*

## Ordinary Regression with Two or More X Variables

StataQuest's **multiple regression** is a straightforward extension of simple regression to models with two or more independent variables. Given a dependent Y variable and independent variables X1, X2, ... Xk, SQ fits the model $Y = b1*X1 + b2*X2 + ... + bk*Xk + c + error$. SQ estimates the slopes (b1, b2, ..., bk) and the constant (c) that best fit your data on the assumption that Y and each X are linearly related and that the effects of the X variables add together to jointly determine Y. The result is called the regression equation.

An illustrative regression equation is $Y = 4.24*X1 - 270.01*X2 + 2500.17$. This equation, computed for sample of 189 births in a hospital survey (*lowbw.dta*, on disk), summarizes the linear-additive relationship between infant birth weight in grams (Y), mother's weight in pounds (X1), and a variable scored 1 if the mother smoked during pregnancy and zero otherwise (X2). The slope b1 is 4.24, the slope b2 is -270.01, and the constant c is 2500.17 grams for the sample data.

In addition to the model fit, SQ's multiple regression computes standard errors, confidence intervals, and other statistics used for inference. It provides **post-regression diagnostics**, including a plot of the model fit, a plot of residuals versus predicted values, and a normal quantile plot of residuals. SQ also allows

---

■ ■ ■ ■ ■ ■

## IN THIS CHAPTER

- ■ **Multiple regression and regression diagnostics**
- ■ **Stepwise regression with forward or backward selection**
- ■ **Using the *fit* command to obtain AV plots and influence statistics**
- ■ **Robust regression**

---

optional forward and backward **stepwise regression**. For advanced users, SQ's *fit* command offers additional regression tools, including **AV plots** (partial-regression leverage plots) and **influence statistics**. SQ's **robust regression** provides one method for coping with the effects of outliers and influential cases.

## Multiple Regression and Regression Diagnostics

Let's use StataQuest's multiple regression to test a multivariate model of average life expectancy in the world's nations. The data are in the *human-dev.dta* file (on disk). The dependent variable is named *life*, average life expectancy in years at birth in 1987. We hypothesize that three different factors contribute to higher average life expectancy: literacy, economic development, and urbanization. The first we'll measure with the variable named *literacy*, the percentage adult literacy rate in 1985. As an indicator of economic development, we'll use the variable named *loggnp* (base-10 logarithm of Gross National Product per capita in U.S. dollars in 1987). And to measure urbanization, we'll use the variable named *urbpop88*, the percentage of the population living in urban areas in 1988. These three variables—*literacy*, *loggnp*, and *urbpop88*—are the independent (X) variables in our model.

The model we want to fit using SQ's multiple regression is *life* = b1*literacy + b2*loggnp + b3*urbpop88 + c + error.

**1** From the *File* menu, choose *Open*.

**2** Enter the name of the file: humandev.

**3** From the *Statistics* menu, choose *Multiple regression*.

**4** **Windows:** Click on the variable named *life* as the dependent variable and on the variables named *literacy*, *loggnp*, and *urbpop88* as the independent variables. Click on the option box to show the diagnostics menu if it is not already selected. Click on the radio button to *Use all selected variables (standard)* if it is not already selected.

**DOS:** Enter the variable named `life` as the dependent variable and the variables named `literacy`, `loggnp`, and `urbpop88` as the independent variables. When prompted, ask for the diagnostics menu and the standard regression.

The filled-out dialog box appears in Figure 18.1.

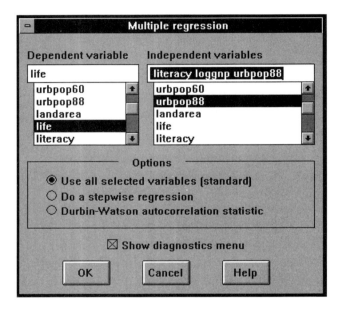

**Figure 18.1** *Dialog Box for StataQuest's Multiple Regression*

Now click on OK to run the multiple regression. The results appear in Figure 18.2.

## Interpreting the Results

Based on the output shown in the "Coef." column of Figure 18.2, the model fit (rounding off) is *life* = .199 * *literacy* + 7.601 * *loggnp* + .016 * *urbpop88* + 24.522. The slope estimates (.199, 7.601, and .016) are called **partial regression coefficients**. Under a causal interpretation, each coefficient estimates the effect of an X variable on Y statistically controlling for the effects of the others.

To illustrate, the coefficient of .199 for *literacy* tells us that average life expectancy increases about one-fifth of a year for each percentage point

```
 Stata Results

    Source |       SS       df       MS              Number of obs =      119
-----------+------------------------------           F(  3,   115) =   207.35
     Model | 11158.3358     3   3719.44528           Prob > F      =   0.0000
  Residual | 2062.82382   115   17.9375985           R-squared     =   0.8440
-----------+------------------------------           Adj R-squared =   0.8399
     Total | 13221.1597   118   112.043726           Root MSE      =   4.2353


-----------------------------------------------------------------------------
      life |     Coef.    Std. Err.     t      P>|t|    95% Conf. Interval]
-----------+-----------------------------------------------------------------
   literacy |  .1994826    .0214711    9.291    0.000   .1569524    .2420127
     loggnp |  7.601359   1.145603    6.635    0.000   5.332141    9.870577
   urbpop88 |  .0162347    .0291063    0.558    0.578  -.0414193    .0738886
      _cons | 24.52196    2.412819   10.163    0.000  19.74263    29.30129
-----------------------------------------------------------------------------
```

**Figure 18.2** *Output of Multiple Regression of Average Life Expectancy on Adult Literacy, Logarithm of Gross National Product, and Percent Urban in 119 Nations*

increase in the adult literacy rate. The coefficient of 7.601 for *loggnp* indicates that average life expectancy increases about 7.6 years for each unit increase in the base-10 logarithm of GNP per capita. The coefficient of .016 for *urbpop88* estimates that average life expectancy increases a rather negligible 16/1000 of a year for each percentage point increase in urbanization. The constant (intercept) of 24.522 years is the expected average life expectancy in the purely theoretical case of zero literacy, zero GNP, and zero urban dwellers.

The partial regression coefficients for *literacy* and *loggnp* are statistically significant at $p < .05$. The coefficient for *urbpop88*, however, is not statistically significant under controls for *literacy* and *loggnp*. This variable probably should be dropped from the model and the regression run again. The Adjusted R-square of .8399 indicates that the current model, which includes *urbpop88*, statistically explains about 84% of the variation in average life expectancy among the 119 nations analyzed.

## Post-Regression Diagnostics

The post-regression diagnostics available following a multiple regression are essentially the same as those following simple regression (see Chapter 17), with one exception. The *Plot fitted model* option after simple regression is replaced here by *Plot actual Y vs. prediction*. The joint effects of multiple X variables are summarized in the single predicted Y value scaled on the bottom axis. No regression **line** is (or can be) drawn. The reason for this change is that in multiple regression there are two or more X variables to plot as opposed to just one X in simple regression. A three-dimensional graph would allow us to plot two X variables and perhaps even a regression **plane**, but with three or more X variables, we'd still have a problem displaying the results.

Figure 18.3 shows the four graphs obtained in succession from the post-regression diagnostics menu following the multiple regression we ran earlier.

The plot of each nation's actual average life expectancy versus its predicted average life expectancy (*_Yhat*) from the equation is shown in the upper left of Figure 18.3. The scatter of data points is fairly tightly clustered and clearly linear, reflecting a good overall model fit. The regression equation is printed at the top of the graph. (Because the full equation is too long to fit within the available space, only the first part is shown with ellipses indicating there's more.)

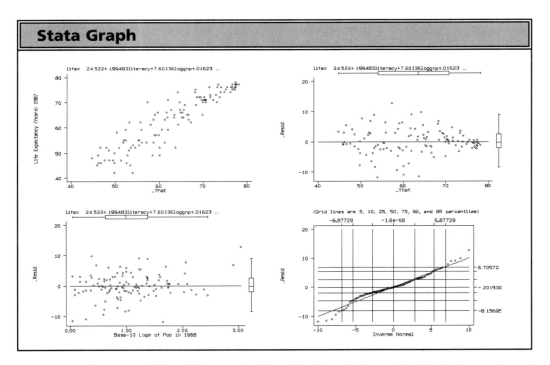

**Figure 18.3** *Four Diagnostic Plots Displayed After a Multiple Regression of Average Life Expectancy on Adult Literacy, Log of GNP, and Percent Urban in 119 Nations*

The standard plot of residuals versus predicted values is displayed in the upper right of Figure 18.3. Overall, the residuals look "healthy" (see Chapter 17), although the diamond-shaped pattern of the scatter suggests some degree of heteroscedasticity.

The normal quantile plot of residuals in the lower right of Figure 18.3 indicates that the residuals are approximately normally distributed. Most of the data points are close to the straight line reflecting normality. A few nations are shown farther out than they "should" be at the low and high ends of the scale, indicating a somewhat heavy-tailed distribution.

Finally, in the lower left of Figure 18.3, the residuals from the equation are plotted against a new X. The new X in this case is *logpop88*, the base-10 logarithm of total 1988 population in millions. There is no apparent relationship between the residuals and *logpop88*, so we are inclined to reject this variable as a candidate for inclusion in our model. (But see the discussion of AV plots that follows.)

## Stepwise Regression with Forward or Backward Selection

SQ's **standard** multiple regression includes all specified independent variables in the model. This option will be pre-selected as the default choice in the dialog box. If you prefer running a **stepwise** multiple regression, select the *Do a stepwise regression* option. Stepwise multiple regression "automates" the selection, keeping, and dropping of X variables from a user-specified variable list.

We believe all analysis should be thoughtful, deliberate, and guided by theory. Thus, we are not great fans of automated regression analysis. Used sparingly and carefully, however, stepwise regression can be a valuable tool for exploratory analysis.

SQ's stepwise procedure has two flavors, forward and backward. **Forward inclusion** starts with the constant and then enters only those X variables into the model that achieve a minimum threshold of statistical significance. Variables are tested and added to the model one at a time, the most significant first. **Backward elimination** starts with all X variables in the model and then drops those that fail to achieve a minimum threshold of statistical significance. Variables are tested and eliminated from the model one at a time, the least significant first.

SQ's stepwise regression uses the t-statistic to test variables for entrance or removal. The significance thresholds are defined by the p-values set in the stepwise regression dialog box. The default "entrance" p-value is .20. The default "removal" p-value is .40. You can change these to other values between

0 and 1 if you wish. To avoid an error message (and to prevent endless recycling of variables in and out of the model), make sure the removal p-value exceeds the entrance p-value. The smaller you set the entrance p-value, the harder it will be for a variable to enter the model. The smaller you set the removal p-value, the harder it will be for a variable to stay in the model.

To demonstrate these two types of stepwise regression, we'll run each of them on the variables we've studied thus far in building a model to explain differences in average life expectancy among nations. In both examples we'll accept the default p-values of .20 for entrance and .40 for removal.

## Example 1: Forward Inclusion

**1**   From the *File* menu, choose *Open*.

**2**   Enter the name of the file: humandev.

**3**   From the *Statistics* menu, choose *Multiple regression*.

**4**   **Windows:** Click on the variable named *life* as the dependent variable and on the variables named *literacy*, *loggnp*, and *urbpop88* as the independent variables. Click on the option box to show the diagnostics menu if it is not already selected. Click on the radio button to *Do a stepwise regression*. In the stepwise regression dialog box, click on the button to select *Forward stepwise procedure*.

   **DOS:** Enter the variable named life as the dependent variable and the variables named literacy, loggnp, and urbpop88 as the independent variables. When prompted, ask for the *Diagnostics* menu and the *Variable selection* procedure. Then choose the *Forward stepwise* procedure.

The results appear in Figure 18.4.

Printed at the top of the regression output table in Figure 18.4 is a step-by-step summary of the forward inclusion procedure. SQ started with an "empty" model (except for the constant), then added the most significant variable *literacy* based on its t-statistic and p-value, then added the next most significant variable *loggnp* based on its t-statistic and p-value, and then stopped and reported the results. The variable *urbpop88* was excluded from the model because the p-value for its t-statistic did not achieve the .20 threshold. Recalling our earlier interpretation of the standard regression results reported in Figure 18.1, we were inclined to drop this variable anyway, and for the same reasons. Compared with the original model, this new one is parsimonious because it statistically explains the same dependent variable with only two X variables rather than three. Note that by excluding the irrelevant *urbpop88* from the model we actually **increased** the Adjusted R-square from .8399 (see Figure 18.2) to .8409.

```
Stata Results

start with empty model
add       literacy    p = 0.0000 <= 0.2000
add       loggnp      p = 0.0000 <= 0.2000

  Source |       SS       df       MS              Number of obs =      119
---------+------------------------------           F(  2,   116) =   312.73
   Model | 11152.7553     2  5576.37764            Prob > F      =   0.0000
Residual | 2068.40439   116  17.8310723            R-squared     =   0.8436
---------+------------------------------           Adj R-squared =   0.8409
   Total | 13221.1597   118  112.043726            Root MSE      =   4.2227

------------------------------------------------------------------------------
    life |    Coef.   Std. Err.      t     P>|t|     [95% Conf. Interval]
---------+--------------------------------------------------------------------
literacy | .2028805  .0205275     9.883   0.000     .1622232    .2435378
  loggnp | 8.025562  .8541897     9.396   0.000     6.333732    9.717392
   _cons | 23.76534  1.989453    11.946   0.000     19.82498    27.7057
------------------------------------------------------------------------------
```

**Figure 18.4** *Output of Stepwise Regression with Forward Inclusion*

## Example 2: Backward Elimination

**1** From the *File* menu, choose *Open*.

**2** Enter the name of the file: humandev.

**3** From the *Statistics* menu, choose *Multiple regression*.

**4** **Windows:** Click on the variable named *life* as the dependent variable and on the variables named *literacy*, *loggnp*, and *urbpop88* as the independent variables. Click on the option box to show the diagnostics menu if it is not already selected. Click on the radio button to *Do a stepwise regression*. In the stepwise regression dialog box, click on the button to select *Backward stepwise procedure*.

**DOS:** Enter the variable named life as the dependent variable and the variables named literacy, loggnp, and urbpop88 as the independent variables. When prompted, ask for the diagnostics menu and the variable selection procedure. Then choose the backward stepwise procedure.

The results appear in Figure 18.5.

As summarized at the top of Figure 18.5, SQ started with the "full" model (including the constant), then eliminated the least significant variable *urbpop88* based on its t-statistic and p-value, and then stopped and reported

**Stata Results**

```
start with full model
remove    urbpop88    p = 0.5781 >  0.4000

  Source |       SS        df       MS              Number of obs =       119
---------+---------------------------------         F( 2,    116) =    312.73
   Model |  11152.7553     2   5576.37764           Prob > F       =    0.0000
Residual |  2068.40439   116   17.8310723           R-squared      =    0.8436
---------+---------------------------------         Adj R-squared  =    0.8409
   Total |  13221.1597   118   112.043726           Root MSE       =    4.2227

-------------------------------------------------------------------------------
    life |    Coef.    Std. Err.     t      P>|t|    [95% Conf. Interval]
---------+---------------------------------------------------------------------
literacy |  .2028805   .0205275    9.883   0.000    .1622232     .2435378
  loggnp |  8.025562   .8541897    9.396   0.000    6.333732     9.717392
   _cons |  23.76534   1.989453   11.946   0.000    19.82498     27.7057
-------------------------------------------------------------------------------
```

**Figure 18.5** *Output of Stepwise Regression with Backward Elimination*

the results. With p-values of .000, the variables *literacy* and *loggnp* easily met the .40 threshold and were permitted to stay in the model. In every other way the regression results reported in Figures 18.4 and 18.5 are identical. Both forward inclusion and backward elimination converged on the same model. (Such convergence will not always occur, so if you're determined to use stepwise regression, it wouldn't hurt to run it both ways.)

## Using the *fit* Command to Obtain AV Plots and Influence Statistics

Although it is not available in the menus, SQ's *fit* command is the key for unlocking a whole new toolbox of graphs and statistics that greatly enhance and extend the standard multiple regression procedure.

The *fit* command by itself is just another way to do regression. Indeed, the *fit* command typed in the Command window will produce exactly the same output as the multiple regression menu command, only without all the nice options and post-regression diagnostics. So why use it?

The answer is that **after** executing the *fit* command you can run an entire series of other commands to produce new kinds of diagnostic graphs, statistics, and tests. These include AV plots (added-variable plots, also known as partial-regression leverage plots), component-plus-residual plots, leverage-versus-squared-residual (L-R) plots, regression specification error (RESET) tests for omitted variables, heteroscedasticity tests, DFBETAS, DFITS, Cook's distance, standardized and studentized residuals, and leverage hats. To learn more about these and other procedures activated by the *fit* command, click on Help in the multiple regression dialog box.

Here we will demonstrate and briefly discuss just two of these procedures, AV plots and Cook's distance. Both are extremely useful tools for detecting outliers and influential cases in multiple regression analysis. AV plots have the added instructional benefit of visually demonstrating to the novice analyst what is really going on when we "statistically control for the effects of other variables."

## AV Plots (a.k.a. Partial-Regression Leverage Plots)

An AV plot gives you a picture of the **partial** regression of a Y variable on an X variable controlling for the other X variables. It consists of a scatter plot of residuals versus residuals with a regression line. The slope of the line is equal to the partial regression coefficient computed by multiple regression. To illustrate, we'll use the *fit* command and AV plots to extend our multiple regression analysis of average life expectancy in nations.

**1**   From the *File* menu, choose *Open*.

**2**   Enter the name of the file: humandev.

**3**   Type fit life literacy loggnp urbpop88 in the Command window and press the ENTER key.

**4**   Type avplots in the Command window and press the ENTER key.

The results appear in Figure 18.6

The upper-left plot in Figure 18.6 shows the partial regression of *life* on *literacy* controlling for *loggnp* and *urbpop88*. The partial regression slope, standard error of the slope, and t-statistic are printed at the top of the plot. These numbers are identical to those reported in the standard multiple regression output. (For example, the slope of the regression line is .199, exactly equal to the partial regression coefficient reported earlier for the variable *literacy* in Figure 18.2.) The left axis measures the **residuals** from regressing *life* on *loggnp* and *urbpop88*. The bottom axis measures the residuals from regressing *literacy* on *loggnp* and *urbpop88*. What the graph shows, in other words, is the relationship between (1) the residual variation in *life* that can't be explained by *loggnp* and *urbpop88*, and (2) the residual variation in *literacy* that can't be explained by *loggnp* and *urbpop88*. Any relationship between these two residuals, therefore, cannot be spuriously caused by *loggnp* and *urbpop88*. This AV plot provides a visual demonstration of what we mean by "statistically controlling for other variables" in the context of multiple regression.

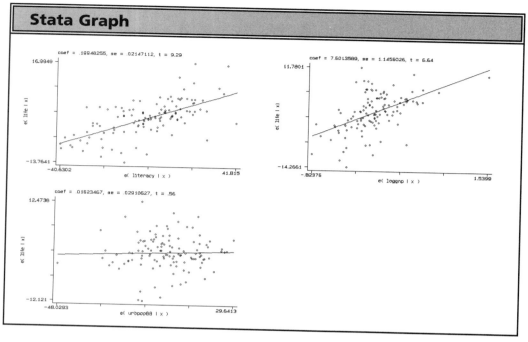

**Figure 18.6** *Added-Variable Plots Showing the Partial Regression of Average Life Expectancy on Each Independent Variable Statistically Controlling for the Effects of the Others. (The slope of each regression line is equal to the partial regression coefficient reported in multiple regression.)*

The other two AV plots in Figure 18.6 have the same design. The upper-right plot shows the partial regression of *life* on *loggnp* controlling for *literacy* and *urbpop88*. The lower-left plot shows the partial regression of *life* on *urbpop88* controlling for *literacy* and *loggnp* (note the flat slope).

Taken together, these three plots graphically decompose the multiple regression into its separate parts. They allow you to **see** the partial regression of the Y variable on each X variable in the model. You can visually assess the goodness of fit and check for non-linearities. You can look for outliers that might be exercising disproportionate influence in estimating a **specific** slope.

To complete this brief illustration of AV plots, suppose you are considering adding *logpop88* to the model as a fourth independent variable. You could do this by adding the variable to the list in multiple regression. But another way to do it and also get a visual display of effects would be to type `avplot logpop88` in the Command window and press the ENTER key. After executing the fit command just once, you can experiment with a whole series of new variables in this fashion. The graphical results are more informative and reliable than those produced by the *Plot residual vs. an X* option in the diagnostics menu. The X in that diagnostic plot is raw; in the AV plot it is residualized.

## Obtaining Influence Statistics Using the *fit* Command

One weakness of ordinary regression is that it tends to track outliers and "wild values" in your data and give these observations too much weight in computing the results. In small data sets, especially, even one wild value can cause ordinary regression to produce completely misleading results. **Robust regression** (see the following section) was developed as an alternative to ordinary regression precisely to deal with such outliers and influential cases.

Although AV plots help, visual **detection of outliers** becomes increasingly difficult if not impossible in multiple regressions involving many independent variables. **Influence statistics** provide tools for detecting outliers and measuring their distorting effects on a multiple regression. These include Cook's distance, DFBETAS, DFITS, leverage hats, and studentized residuals. The logic underlying most influence statistics is to run a regression, remove an observation, and run it again. The degree of influence assigned to an observation is proportional to the impact of its removal on the overall regression results. (For an excellent discussion, see Bollen and Jackman, 1990.) SQ's *fit* command allows you to compute all of these and other influence statistics.

We'll illustrate how to use influence statistics by computing **Cook's distance** ($D_i$) for each observation following a multiple regression of *life* on *literacy*, *loggnp*, and *urbpop88*. The Cook's distance statistic measures the influence of each observation on the overall regression. Observations that have values of $D_i$ exceeding 1.0 are considered extremely "influential." (Some analysts use a lower cutoff value set by the formula 4/n, where n is the number of observations.)

**1** From the *File* menu, choose *Open*.

**2** Enter the name of the file: `humandev`.

**3** Type `fit life literacy loggnp urbpop88` in the Command window and press the ENTER key

**4** Type `fpredict cooksd, cooksd` in the Command window and press the ENTER key.

**5** Type `sort cooksd` in the Command window and press the ENTER key.

**6** Type `list name cooksd` in the Command window and press the ENTER key.

A partial display of the results appears in Figure 18.7

Figure 18.7 reports the name and Cook's distance statistic of the six nations having a $D_i$ exceeding the lower cutoff of 4/119 = .0336. No observation exceeds

**Stata Results**

|      | name      | cooksd   |
|------|-----------|----------|
| 114. | Morocco   | .040177  |
| 115. | Gabon     | .0469409 |
| 116. | Sri Lanka | .0709313 |
| 117. | China     | .0759609 |
| 118. | Lao PDR   | .1026955 |
| 119. | Ethiopia  | .1124589 |

**Figure 18.7** *Nations with the Highest Cook's Distance Following a Multiple Regression of Average Life Expectancy on Adult Literacy, Log of GNP, and Percent Urban in 119 Nations*

the higher cutoff of 1.0. Although these six nations might require special attention, it is clear by this measure that there are no extremely wild values in the data to worry about. Thus we can be more confident in the reliability of our multiple regression results.

## Robust Regression

Ordinary regression methods perform well in a wide variety of analytical situations. But they have an important weakness: they tend to "track" outliers, giving unusual data points disproportionate influence over our conclusions. **Robust regression** refers to a class of regression-type methods designed to overcome this weakness of ordinary regression.

Many varieties of robust regression exist, each with its own advantages. StataQuest's robust regression works by giving each observation a **weight** (w), some number in the range 0 <= w <= 1. Regression residuals tell us how much an observation deviates from the general pattern of the data. StataQuest assigns lower weights to the most deviant observations, those with the largest residuals. Given lower weights, these observations then have less effect on the robust regression. SQ actually performs a series of weighted regressions, each time calculating residuals, assigning weights according to one of two mathematical "weighting functions," and then repeating the regression with these new weights. This process stops when no further improvement occurs. By that time, severe outliers might have weights as low as zero—indicating that the robust regression decided to ignore them entirely. (For technical details on the Huber weight function and Tukey biweight function used in this particular robust regression method, see Hamilton, 1992, 1993.)

---

■ **TIP 21** ■ ■ ■ ■ ■ ■ ■ ■ ■ ■ ■ ■ ■ ■ ■

### The Durbin-Watson Statistic

In SQ's multiple regression dialog box you can select the option to compute the Durbin-Watson statistic to test for autocorrelated errors. Autocorrelation is correlation between values of the same variable across different observations. Autocorrelated errors violate ordinary least-squares (OLS) regression assumptions, reduce the efficiency of OLS, and bias estimates of standard errors. Autocorrelation is particularly a problem in time-series analysis. The Durbin-Watson statistic is used to test for autocorrelated errors by examining the residuals of sample data. For a clear discussion of when and how to use this statistic and also of its limitations, see Hamilton, 1992: 118–121.

---

SQ's robust regression works particularly well when the **dependent** variable exhibits occasional "wild" values as a result of measurement error, extraneous factors, or the inclusion of observations that really don't belong in our sample. In such instances, downweighting the far-out observations makes good sense, and SQ accomplishes it in a smooth, nonarbitrary fashion. On the other hand, robust regression may work no better than ordinary regression when the **independent** variables have wild values. That situation presents a different theoretical and statistical problem, calling for a different kind of "robust regression." Although SQ's (or any other) robust regression method is by no means foolproof, it will often be safer than ordinary (OLS) regression when used by inexperienced data analysts.

When performing an important regression, try it both ways, as an ordinary regression and as a robust regression. Compare the resulting coefficients and standard errors. If they appear similar, report the more widely understood OLS regression results. If they appear different, investigate further to find out why. The difference warns you that outliers may be disproportionately influencing the ordinary regression results.

We'll briefly illustrate SQ's robust regression procedure by analyzing data on marriage rates and age distributions contained in the *states3.dta* file (on disk). The Y variable of interest is *marrate*, the number of marriages per one thousand persons in 1992. The X variable is *pct1824*, the percentage of population between the ages of 18 and 24 years old in 1993.

As you can check for yourself, a simple (OLS) regression of *marrate* on *pct1824* yields the equation Y = 46.278 − 3.519X. The standard error of the slope is 2.217 with a t-ratio of −1.587 (p < .119). The negative slope of −3.519

indicates an inverse relationship: the higher the percentage in the 18–24 age group, the lower the marriage rate. This is a surprising result; one would expect the relationship to be positive, not negative. We decide to run a robust regression to compare with the OLS regression results.

**1** From the *File* menu, choose *Open*.

**2** Enter the name of the file: `states3`.

**3** From the *Statistics* menu, choose *Robust regression*.

**4** **Windows:** Click on the variable named *marrate* as the dependent variable and on the variable named *pct1824* as the independent variable.

**DOS:** Enter the variable named `marrate` as the dependent variable and the variable named `pct1824` as the independent variable.

The robust regression model fit is $Y = 3.147 + .569X$, indicating a positive relationship between *marrate* and *pct1824*. The standard error of the slope is .363 with a t-ratio of 1.569 ($p < .123$). These results are completely contrary to those obtained using ordinary regression. They warn us that one or more outliers might be disproportionately influencing the OLS estimates. The likely culprit is Nevada, which has a marriage rate of 86.1 marriages per thousand persons—much higher than the next highest rate (for Arkansas) of 15.6. A good strategy would be to omit Nevada from the data and run the OLS regression again.

## ■ EXERCISES

**18.1** Open the *states2.dta* file (on disk), which contains various data on the 50 American states. The dependent variable of interest is *toxics*, which measures the toxics released per capita in pounds in 1990.

    a. Run a standard multiple regression of toxics on the following independent variables: *metro*, *miles*, *btu*, *waste*, *farms*, *mines*, and *mftr*. What conclusions do you draw? Based on the results, which variables would you keep in the model, which would you remove? Why?

    b. Run the same multiple regression using stepwise forward inclusion. Compare the results with those obtained in (a).

    c. Run the same multiple regression using stepwise backward elimination. Compare the results with those obtained in (a) and (b).

**18.2** Open the *lowbw.dta* file (on disk), which contains data on a sample of 189 births drawn from a hospital survey. Use multiple regression to study the relationship between infant birth weight in grams and mother's weight, smoking habits, and number of visits to a physician.

**18.3** Open the *states3.dta* file (on disk), which contains various data on the 50 American states. Use multiple regression to study the relationship between 1990 abortion rates and the following independent variables: percentage of the population 16–19 years who were high school dropouts in 1990, per capita income in 1993, percentage of the population 25 years or older who have at least a BA degree, and percentage of the 1990 population who were Christian adherents.

**18.4** (Advanced) Using SQ's *fit* command, extend the analysis performed in Exercise 18.3 by running AV plots for all independent variables. Also compute Cook's distance measures for each state following the regression. What additional insights are gained from this analysis? In particular, are any states identified as outliers or influential cases in determining the regression results?

**18.5** (Advanced) Open the *states2.dta* file (on disk), which contains various data for the 50 American states. The dependent variable of interest is named *green*, which measures each state's greenhouse omissions per capita in pounds in 1990. The independent variables include *metro* (percentage of population living in metropolitan areas in 1990), *btu* (energy consumed per capita in 1990 in millions of British thermal units), and *mines* (percentage of Gross State Product generated by the mining industry in 1989). Run a multiple regression of *green* on *metro*, *btu*, and *mines*. Then run a robust regression of *green* on the same independent variables. Compare the results. What does this comparison suggest regarding the possible presence of influential outliers in the data? What statistical and graphical tools might you use to detect such outliers, if they exist?

**18.6** (Advanced) Open the *robust.dta* file (on disk), which contains made-up data designed to illustrate differences between OLS and robust regression.

  a. Use simple regression to regress y1 on x1, y2 on x2, y3 on x3, and y4 on x4. After each simple regression, note the slope estimate and standard error of the slope, and then plot the model fit.

  b. Use robust regression to regress y1 on x1, y2 on x2, y3 on x3, and y4 on x4. After each robust regression, note the slope estimate and standard error of the slope, and then plot the model fit.

  c. Compare the results obtained from (a) and (b). Which regression procedure, OLS or robust, seems most appropriate to use in each of the four analytical situations. Why?

# 19

# Logistic Regression
## (Windows Only)

## What is Logistic Regression?

Logistic regression (also called logit analysis) is a statistical method for analyzing a categorical dependent (Y) variable as a function of one or more independent (X) variables. It can be used to analyze dichotomous Y variables (two values) and polytomous Y variables (three or more values). As with standard multiple regression, logistic regression can include any number of X variables as predictors. Hamilton (1992), Hosmer and Lemeshow (1989) and Hanushek and Jackson (1977) all provide clear introductions to this method.

SQ's logistic regression procedure is limited to the analysis of dichotomous Y variables and can include only one X variable at a time. Despite these restrictions, it is a welcome addition to SQ's package. It can be used by students to explore a powerful analytical tool that is usually introduced only in intermediate or advanced courses.

Beyond its instructional value, SQ's logistic regression procedure is the tool of choice for analyzing two-variable relationships in which the Y variable is dichotomous and the X variable is continuous. Dichotomous dependent variables are the focus of much research. Examples include voting (voted or didn't vote); marital status (married or not married); employment (employed

■ ■ ■ ■ ■ ■ ■
## IN THIS CHAPTER

■ **The logistic regression model and assumptions**

■ **Using SQ's logistic regression: an example**

■ **Graphing the results**

or unemployed); smoking (smoker or nonsmoker); birth weight (normal birth weight or low birth weight). For various reasons (see Hamilton, 1992; Hosmer and Lemeshow, 1989), standard OLS regression methods are inappropriate for analyzing dichotomous Y variables of this sort. Cross-tabulation works well enough in the bivariate case. This method becomes cumbersome and limited, however, when two or more X variables are involved or when the X variables are continuous rather than categorical.

This chapter will (1) briefly discuss the mathematical model and assumptions underlying logistic regression, (2) demonstrate SQ's logistic regression procedure on a real problem, (3) illustrate and interpret the two different kinds of output (coefficient estimates and odd-ratios) one can choose when using this method, and (4) show how to use graphs as visual aids in making sense of the results.

## The Logistic Regression Model and Assumptions

In the logistic regression model, observations on a dichotomous Y variable are scored as 1 ("success") or 0 ("failure"). For example, if Y is voting turnout, the researcher might score voters as 1 and nonvoters as 0. These binary (0 or 1) data are assumed to be the observed outcomes of an unobserved underlying probability distribution. It is true that people either vote or don't vote; they're either married or unmarried. But the probability that a person will vote or be married can vary on a continuum that ranges theoretically from zero (impossible) to 1.0 (certainty). The higher the probability, the more likely Y = 1; the lower the probability, the more likely Y = 0.

A typical hypothesis is that these assumed-to-exist probabilities are functionally related in some way to an X variable. For example, one might hypothesize that the probability of turning out to vote increases with age. Logistic regression is used to analyze the data to estimate the probability that Y = 1 for different values of X on the assumption that Y and X are functionally related in the following way.

$$\text{Prob}\{Y=1\} = e^{\beta_0 + \beta_1 x} / [1 + e^{\beta_0 + \beta_1 x}], \tag{1}$$

where Prob{Y=1} is the expected value of the probability that Y = 1 conditional on X; $e$ is 2.71828..., the base of the natural logarithm; X is the independent variable; and $\beta_0$ and $\beta_1$ are unknown parameters to be estimated from the data using maximum likelihood methods. These estimates of $\beta_0$ and $\beta_1$ are known as the model fit. An equivalent formula that is easier to use in computing conditional probabilities from model fits is:

$$\text{Prob}\{Y=1\} = 1 \ / \ [ \ 1 + e^{-L} \ ], \tag{2}$$

where the exponent $L$ (also known as the logit) is $\beta_0 + \beta_1$ X. This mathematical model assumes a nonlinear "S"-shaped relationship between Prob{Y=1} and X, with predicted probabilities bounded between 0 and 1. This is in contrast with the linear relationship assumed in simple OLS regression. One reason OLS regression does not work on dichotomous Y variables is that the assumed linear relationship can produce estimated probabilities less than 0 or greater than 1 for realistic values of X.

Given data on Y and X for a set of observations, SQ's logistic regression program iteratively searches for and finds the best-fitting values of $\beta_0$ and $\beta_1$ that make the observed data "most likely" to occur on the assumption that the logistic model is true. It does this by maximizing what is called a "log likelihood" function, which will not be defined here. (For a clear discussion, see Hosmer and Lemeshow, 1989.) These computed **log likelihoods** are an important part of the output of logistic regression.

An understanding of odds, log of the odds (logit), and odds-ratios will help in interpreting logistic regression results.

## Odds

The **odds** that Y=1 are calculated as Prob{Y=1} divided by Prob{Y=0}. Letting P = Prob{Y=1} and 1-P = Prob{Y=0}, the odds are defined as P/(1–P). For example, if the probability is .80 that an individual will vote, the probability of not voting is 1–P = .20, and the odds of voting are .80/.20 = 4.

## Log of the Odds (Logit)

The **log of the odds** that Y=1 is defined as $\log_e$ [P/(1–P)]. This is sometimes called the logit transform or just **logit**. Logits can vary from minus to plus infinity. Continuing the example, if the odds of voting are 4, the log of the odds or logit is 1.386. The linear relationship underlying logistic regression—and the one that links it to an entire family of linear models including OLS regression—is the following:

$$\log_e \ [\text{P}/\,(\text{1-P})\,] = \beta_0 + \beta_1 \ \text{X} \tag{3}$$

It helps to know this equation because the coefficient estimates reported in the logistic regression output have a direct linear interpretation only if the logit is understood as the dependent variable. For example, if the estimate of $\beta_1$ is .500, that implies the log of the odds or logit will increase by half a point for each unit increase in X. It does **not** mean that Prob{Y=1} will increase by half a point. To find out how much Prob{Y=1} changes (nonlinearly) with a change in X, one must insert the $\beta_0$ and $\beta_1$ coefficient estimates into one of the two formulas shown above for Prob{Y=1} and then try out different values of X. As will be illustrated later, graphs are particularly helpful in visualizing how Prob{Y=1} varies with changes in X. (For a useful guide to interpreting logistic regression output, see Liao, 1994.)

## Odds-Ratios

SQ's logistic regression procedure can report results in two different ways. The first is a standard report of coefficient estimates, standard errors, and related information. As just noted, these results must be interpreted carefully while keeping in mind the nonlinear form of the model being tested. The second is a report of the **odds-ratio** (often symbolized as $\psi$), its standard error, and its 95% confidence interval. Odds-ratios are easier to interpret than coefficient estimates, especially if X is also a dichotomous variable scored 0 or 1. SQ's logistic regression will report the odds-ratio rather than coefficient estimates if you click on the "display odds-ratio" option in the dialog box.

### Odds-Ratios with Dichotomous X

In the case of a dichotomous X, the odds-ratio is the ratio of the odds that Y=1 when X=1 to the odds that Y=1 when X=0. To illustrate with data collected in Hosmer and Lemeshow's study of low birth weight babies (*lowbw.dta*), here is a cross-tabulation of baby's birth weight (normal or low) by mother's report of smoking during pregnancy (no or yes).

```
Low birth| Smoke during pregnancy? 1=yes
  wt <2500|
grams 1=low|        No        Yes |     Total
-----------+----------------------+----------
   Normal |        86         44 |       130
          |     74.78      59.46 |     68.78
-----------+----------------------+----------
      Low |        29         30 |        59
          |     25.22      40.54 |     31.22
-----------+----------------------+----------
    Total|        115         74 |       189
          |    100.00     100.00 |    100.00

        Pearson chi2(1) =   4.9237   Pr = 0.026
```

Based on these results, the odds that smokers (X=1) have low birth weight babies (Y=1) are .4054/.5946=.6818, and the odds that nonsmokers (X=0) have low birth weight babies (Y=1) are .2522/.7478= .3373. The odds-ratio for smokers having low birth weight babies relative to non-smokers is

$$\psi \ = \ .6818/.3373 \ = \ 2.02$$

Odds-ratios can also be computed directly using the cross-products of table cell counts—for example, $\psi = 30/44$ divided by $29/86 = (30 \times 86)/(44 \times 29) = 2.02$.

In the example, the odds-ratio of 2.02 tells us that smokers are more than twice as likely as nonsmokers to deliver low birth weight babies. If smokers were no more likely than nonsmokers to deliver low birth weight babies, the odds-ratio would be 1.00. If they were actually less likely than nonsmokers to do so, the odds-ratio would be less than 1.00.

### Odds-Ratios with Continuous X

Odds-ratios can be a bit more difficult to interpret if the X variable has more than two values. To illustrate, an odds-ratio of .986 is reported for the relationship between low birth weight babies (Y) and weight of mother in pounds (X). Mother's weight is a continuous variable that ranges from 80 to 250 pounds in the data. The odds-ratio of .986 tells us that mothers of a certain weight are .986 times as likely as those who weigh one pound less to deliver low birth weight babies. This is useful information, but a more informative statistic might be an odds-ratio showing the effects of a 10-pound difference in weight. This can be done simply by generating a new variable (call it X2) equal to mother's weight divided by 10 and then obtaining an odds-ratio for X2 rather than X. In the example, that new odds-ratio is .869, indicating that mothers of a certain weight are .869 times as likely as those who weigh ten pounds less to deliver low birth weight babies.

# Using SQ's Logistic Regression: An Example

This example will use data from Hosmer and Lemeshow's study of risk factors associated with low infant birth weight (use *lowbw.dta*). The dependent variable is *low* (1 = low birth weight, 0 = normal birth weight). The independent variables of interest are *smoke* (1 = mother smoked during pregnancy, 0 = did not smoke) and *lwt* (mother's weight in pounds). Because the dependent variable *low* is dichotomous and has been scored 0 or 1, SQ's logistic regression procedure is appropriate to use to test the hypotheses that (1) mothers who smoke are more likely than mothers who don't to deliver low birth weight babies, and (2) the likelihood of low birth weight deliveries decreases with increasing mother's weight.

To illustrate the two different kinds of output produced by SQ's logistic regression procedure, we'll use odds-ratios to test the first hypothesis and coefficient estimates to test the second.

### Low Birth Weight and Smoking: Odds-Ratios

**1**  From the *File* menu, choose *Open*.

**2**  Enter the name of the file: lowbw.

**3**  From the *Statistics* menu, choose *Logistic regression*.

**4**  **Windows:** For the dependent variable, click on *low*. The independent variable should be *smoke*. Under "Options," click on the box to request odds-ratios. Click on OK.

The dialog box is shown in Figure 19.1; the results of the procedure are in Figure 19.2.

The output display has three basic parts: (1) log likelihoods for each iteration of the maximum likelihood procedure; (2) hypothesis tests and goodness-of-fit statistics, including results of the log likelihood test and the Pseudo-$R^2$; (3) the estimated odds-ratio ($\psi$) with its standard error and 95% confidence interval.

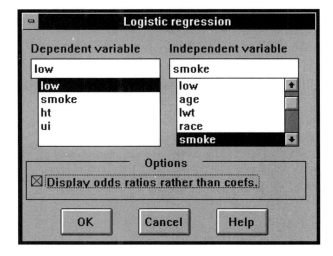

**Figure 19.1** *Dialog Box for Logistic Regression*

```
 ┌─────────────────────────────────────────────────────────────────┐
 │ Stata Results                                                     │
 ├─────────────────────────────────────────────────────────────────┤
 │                                                                   │
 │  . logit low smoke, or                                            │
 │                                                                   │
 │  Iteration 0:   Log Likelihood =  -117.336                        │
 │  Iteration 1:   Log Likelihood = -114.9123                        │
 │  Iteration 2:   Log Likelihood = -114.9023                        │
 │                                                                   │
 │  Logit Estimates                        Number of obs =      189  │
 │                                         chi2(1)        =     4.87  │
 │                                         Prob > chi2    =   0.0274  │
 │  Log Likelihood =  -114.9023            Pseudo R2      =   0.0207  │
 │                                                                   │
 │  -------------------------------------------------------------    │
 │   low | Odds Ratio  Std. Err.    z    P>|z|  [95% Conf. Interval]  │
 │  -----+-------------------------------------------------------    │
 │ smoke |   2.021944  .6462912  2.203  0.028   1.080668   3.783083  │
 │  -------------------------------------------------------------    │
 │                                                                   │
 └─────────────────────────────────────────────────────────────────┘
```

**Figure 19.2** *Logistic Regression Output (Odds-Ratio)*

### Log Likelihoods
The initial log likelihood (iteration 0) is −117.336. The final log likelihood (iteration 2) is −114.9023. Because the log likelihood test statistic (see next paragraph) uses both of these numbers, they should be included in any report of logistic regression results.

### Hypothesis Testing and Goodness-of-Fit
The chi-squared of 4.87 is the log likelihood test statistic, computed as −2 × [initial log likelihood − final log likelihood]. The initial log likelihood is computed for a model that includes just the dependent variable. The final log likelihood gives the maximum likelihood solution with the independent variable included. The greater the difference between initial and final log likelihoods, the greater the improvement in model fit as a result of including the independent variable.

In this example, the difference is [−117.336 − (−114.9023)] = −2.4337. Multiply this result by −2 and the log likelihood statistic is 4.8674. Under the null hypothesis that $\beta_1 = 0$, this statistic follows the chi-squared distribution with one degree of freedom for each parameter estimated except for the constant term. The reported p-value is .0274, indicating that the relationship between *low* and *smoke* is statistically significant at p < .05.

The Pseudo-$R^2$ is .0207. This goodness-of-fit statistic is calculated as $1 - (L_f/L_i)$, where $L_f$ is the final log likelihood and $L_i$ is the initial log likelihood. (For an alternative definition of Pseudo-$R^2$, see Hamilton, 1992). It is analogous to the $R^2$ of OLS regression, although it does not have the same variance-explained interpretation. It should be used only as a rough measure of the improvement in goodness-in-fit in comparisons of models with the same dependent variable. Alternative goodness-of-fit measures are discussed in Hosmer and Lemeshow (1989).

### Estimated Odds-Ratio

The estimated odds-ratio is 2.02, the same number arrived at earlier using the cross-tab counts. This odds-ratio indicates that mothers who smoke are about twice as likely as those who don't to deliver low birth weight babies. The standard error of this estimate is .6462912 and is statistically significant at p < .05. (The value of 0.028 under P>|z| means that only 28 out of a 1000 independent samples of this size would have an odds-ratio this different from 1.0 under the null hypothesis that the population odds-ratio is 1.0.) The 95% confidence interval is (1.080688, 3.783083), meaning that we can be 95% sure that the population odds-ratio falls between 1.08 and 3.78. This interval does not include 1.0, which would indicate no correlation between birth weight and smoking.

**Note:** As the reader can check by repeating the same analysis without the odds-ratio option, the estimated $\beta_1$ is .7040592. For **dichotomous** X variables as used here, a convenient formula linking the coefficient estimate $\beta_1$ and the odds-ratio $\psi$ is: $\psi = e^{\beta 1}$. In this example, $e^{.7040592} = 2.02$.

## Low Birth Weight and Mother's Weight: Coefficient Estimates

**1** From the *File* menu, choose *Open*.

**2** Enter the name of the file: `lowbw`.

**3** From the *Statistics* menu, choose *Logistic regression*.

**4** **Windows:** For the dependent variable, click on *low*. The independent variable should be *lwt*. Under "Options," do **not** click on the box to request odds-ratios. Click on OK.

The results of the procedure are in Figure 19.3.

The logistic regression output reports estimates of the $\beta_0$ and $\beta_1$ coefficients along with their standard errors and confidence intervals.

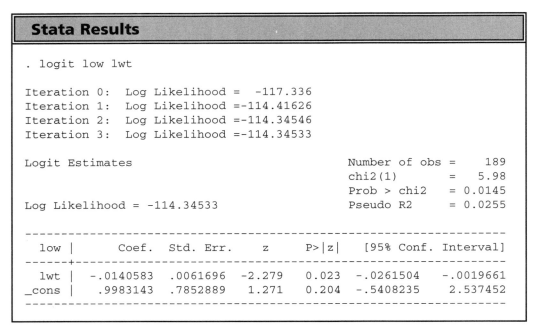

**Figure 19.3** *Logistic Regression Output (Coefficient Estimates)*

### Log Likelihoods

The initial log likelihood (iteration 0) is -117.336. The final log likelihood (iteration 3) is -114.34533.

### Hypothesis Testing and Goodness-of-Fit

The chi-squared of 5.98 is the log likelihood test statistic. Under the null hypothesis that $\beta_1 = 0$, this statistic follows the chi-squared distribution with one degree of freedom. The reported p-value is .0145, indicating that the relationship between *low* and *lwt* is statistically significant at $p < .05$. The Pseudo-$R^2$ is .0255.

### Coefficient Estimates

The estimate of $\beta_1$ for the variable *lwt* is –.0140583, the minus sign indicating a negative relationship between the probability of low infant birth weight and mother's weight. The standard error of this estimate is .0061696 and is statistically significant at $p < .05$. The confidence interval of (–.0261504, –.0019661), means that we can be 95% sure that $\beta_1$ falls between –.026 and –.002 in the survey population. The estimate of $\beta_0$ is .9983143, which is shown for *_cons* (the constant).

Based on these estimates of $\beta_0$ and $\beta_1$, and using equation 3 on page 270, the model fit for predicting the log of the odds or logit as a function of mother's weight (X) is:

$$\text{Log}_e \ (P/1\text{-}P) \ = \ .998 \ - \ .014X$$

Thus, for every one-pound increase in mother's weight, the **log of the odds** of a low birth weight delivery decreases by .014 of a point.

This is useful information, but what we'd really like to know is how the **probability** of a low birth weight delivery changes with increasing mother's weight. To find that out, we need to substitute the estimates of $\beta_0$ and $\beta_1$ into equation 1 or equation 2 on page 270. Using equation 2, the model fit for the expected value of the probability of low birth weight delivery as a function of mother's weight (X) is the following:

$$\text{Prob}\{Y=1\} \ = \ 1 \ / \ [ \ 1 \ + \ e^{-L} \ ] \ = \ 1 \ / \ [1 \ + \ e^{-(.998 \ - \ .014X)}]$$

Skipping the math, this equation tells us that there is a **nonlinear** relationship between the probability of low birth weight and mother's weight. Specifically, the **effect** of mother's weight on the probability of low birth weight delivery is conditional on **level** of mother's weight. In this particular example, the heavier the mother, the less effect each additional pound will have in reducing the probability of a low birth weight delivery.

# Graphing the Results

Although there are different mathematical ways to summarize and interpret the effects of an X variable on predicted probabilities (see Liao, 1994), the clearest and most intuitive approach is to graph them. To produce a graph, we first need to use the coefficient estimates and equation 2 on page 270 to compute predicted values of Prob{Y=1} for each X value in the data. To continue our example, the predicted probability of a low birth weight delivery for an 80-pound mother is $1 \ / \ [1 \ + \ e^{-(.998 \ - \ .014*80)} \ ] \ = \ .470$. The predicted probability for a 250-pound mother is $1 \ / \ [1 \ + \ e^{-(.998 \ - \ .014*250)} \ ] \ = \ .076$.

These calculations can be made by using your own or SQ's calculator. You can also use SQ's display command. For example, in the Command window (Windows), enter the following:

```
display 1/(1 + exp(-(.998 - .014*250)))
```

(As an exercise, try this display command to compute the probability for a 300-pound mother. Try it for a 400-pound mother. Notice that the predicted probabilities become lower with increasing mother's weight, but do not fall below zero. Also see Exercise 19.2 at the end of this chapter.) The easiest way to compute predicted probabilities from the model fit is to use SQ's predict command immediately after the last-run logistic regression. First, run the logistic regression procedure. Then enter the command:

```
predict phat
```

where *phat* is the name SQ will assign to the new variable storing the computed probabilities for each observation in the data set. (You can choose some other variable name if you want. Be sure you have enough room within the spreadsheet limits to store this new variable, which will be added to your variable list.)

The next step is to plot these predicted probabilities against the X variable *lwt*. Select *Scatter Plots* and then *Plot Y vs. X* from the *Graphs* menu. Select *phat* as the Y-axis variable; *lwt* is the X-axis variable. The result is shown in Figure 19.4.

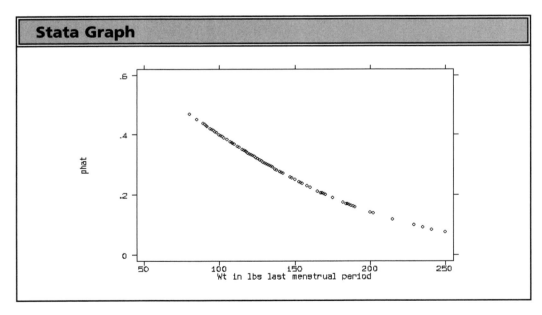

**Figure 19.4** *Plot of Predicted Probabilities vs. X Variable*

As can be seen in Figure 19.4, the line of dots begins to curve and flatten as mother's weight increases. This visually illustrates the changing slope of the (nonlinear) relationship between mother's weight and the predicted probability of low birth weight delivery.

# ■ EXERCISES

**19.1** Open the *vote.dta* dataset, which has information on reported voting turnout in the 1992 presidential election (*vote92*), education level (*educ*), age (*age*) and other variables for a subsample of 598 adult respondents interviewed in a national sample survey. After checking to be sure that the dichotomous dependent variable *vote92* is scored 0 or 1, do the following:

   a. Use logistic regression with odd-ratios to test the hypothesis that voting turnout increases with higher education.

   b. Use logistic regression with coefficient estimates to test the hypothesis that voting turnout increases with age.

   c. Using the estimates obtained in (b), plot the predicted probabilities of voting against age.

   d. Select another independent variable from the data set, choose the appropriate kind of output (odds-ratios or coefficient estimates), run a logistic regression, produce a graph (if appropriate), and interpret the results.

**19.2** Open the *lowbw.dta* dataset. Do the following:

   a. As was done in the text, use logistic regression to obtain estimates of $\beta_0$ and $\beta_1$ for the relationship between *lowbw* and *lwt*. Use those estimates and equation (2) to compute the predicted probability of a low birth weight delivery for a 300-pound mother. Do the same for a 400-pound mother.

   b. Select *Statistics/Simple Regression*, regress *lowbw* (Y) on *lwt* (X), and then use the estimates of $\beta_0$ and $\beta_1$ in the equation $Y = \beta_0 + \beta_1 X$ to compute the predicted probabilities of a low birth weight delivery for a 300-pound mother. Do the same for a 400-pound mother. Compare these results with those obtained in (a). What does this comparison suggest regarding a choice between logistic regression and simple OLS regression as the preferred method for problems of this sort (that is, for problems that involve dichotomous dependent variables)?

# 20

# *Analysis of Variance (ANOVA)*

## StataQuest's ANOVA Command

StataQuest's ANOVA command can be used to run different analyses of variance models, including balanced designs, unbalanced designs, repeated measures ANOVA, designs with missing cells, and designs with interaction terms. The essence of analysis of variance is obtaining a statistical test of the differences among the average performance on a dependent variable among various subgroups. Most often, you are testing the difference between one or more experimental or treatment groups, which have experienced some level of a treatment in an experiment, and one or more control groups, which have been given a placebo or have not experienced the treatment. The goal is to determine whether the groups are statistically significantly different given their means, standard deviations, and sample sizes.

## Nonparametric Analysis of Variance

This test is the same as the Kruskal-Wallis test of the equality of populations, covered in Chapter 14. Here we use the test as a one-way analysis of variance test, one-way meaning that one independent variable will be tested. The same

■ ■ ■ ■ ■ ■ ■

# IN THIS CHAPTER

- **Nonparametric analysis of variance**
- **One-way ANOVA**
- **Two-way ANOVA**
- **Repeated measures ANOVA**
- **N-factor ANOVA & ANOCOVA**

---

caveats that apply in Chapter 14 apply here also, namely, that we assume that each of the categories of the categorical variable has at least five observations, although these do not have to be equally sized, and the observations within categories should be continuous. The test relies on ranks and does not make the same assumptions as regular ANOVA (particularly the assumption of normally distributed data), so it can be used with data that are ordinal and not interval or ratio level. The test relies on the rank of each observation on the independent variable and not on the mean and standard deviation. The null hypothesis is usually stated that the population distributions are identical.

Our data set is from a study of the effects of vitamin C on the survival times of terminal cancer patients, conducted by Dr. Linus Pauling and associates (Cameron and Pauling, 1976, 1978, Andrews and Herzberg, 1985). We have excerpted 30 observations from their larger dataset, 13 from the patients with stomach cancer and 17 from the patients with cancer of the colon. For each patient, we have two measures of the length of time the patient survived. The first, in variable *a*, is the number of days the patient survived "from the date of first hospital attendance for the cancer that reached the terminal stage"; the second, in variable *b*, is the number of days the patient survived "measured from the dates of untreatability" (Andrews and Herzberg, 1985, 206).

We will run the nonparametric ANOVA procedure with both *a* and *b* as dependent variables. The independent variable is *ctrl*, a 1/0 dichotomous variable to indicate whether the patient received vitamin C.

**1** From the *File* menu, choose *Open*.

**2** Enter the name of the file: `pauling`.

**3** From the *Statistics* menu, choose *ANOVA* and then *1-way nonparametric*.

**4** **Windows:** For the data variable, click on *a*. The group variable should be *ctrl*. Click on OK.

   **DOS:** In answer to the questions, the dependent variable should be a, and the variable identifying the groups should be `ctrl`.

**5** Repeat the analysis for the second dependent variable, *b*.

```
  Stata Results

. kwallis a, by(ctrl)

Test: Equality of populations (Kruskal-Wallis Test)

       ctrl        _Obs     _RankSum
    Treatmt          30       986.50
    Controls         30       843.50

chi-squared =     1.117 with 1 d.f.
probability =     0.2905

. kwallis b, by(ctrl)

Test: Equality of populations (Kruskal-Wallis Test)

       ctrl        _Obs     _RankSum
    Treatmt          30      1257.50
    Controls         30       572.50

chi-squared =     25.641 with 1 d.f.
probability =     0.0001
```

**Figure 20.1** *Nonparametric ANOVA, Effectiveness of Vitamin C Data*

The results are shown in Figure 20.1.

Figure 20.1 shows that we have 30 observations (_Obs) in the treatment group and 30 in the control group. The "_Ranksum" information is used in computing the test. We receive a chi-squared value of 1.117 for *a* and 25.641 with *b*, both with 1 degree of freedom. The number of degrees of freedom for this test is one less than the number of groups; we had two groups, the treatment group and the control group. The probability of 0.0001 for *b* indicates that we have 1 chance in 10,000 that we could have received this result by chance were the null hypothesis that the two groups are identical true.

## One-Way ANOVA

One-way ANOVA tests the null hypothesis that the mean of the dependent variable is the same for each category of a single independent variable. In contrast with nonparametric ANOVA, one-way ANOVA relies on the mean and standard deviation to perform this test and consequently assumes that the data points are distributed normally around each subgroup mean, that each subgroup has the same variance, and that each case in the sample is independent of the others.

We will use the *pauling.dta* data set to do a one-way analyses, but instead of doing two analyses as we did with the nonparametric test, we will perform the analysis on *b* to save space. The reader can do the analysis on *a* to see whether the results are different from the nonparametric case.

**1**  From the *File* menu, choose *Open*.

**2**  Enter the name of the file: pauling.

**3**  From the *Statistics* menu, choose *ANOVA* and then *One-way*.

**4**  **Windows:** For the data variable, click on *b*. The group variable should be *ctrl*. Click on OK.

> **DOS:** In answer to the questions, the dependent variable should be b, and the variable identifying the groups should be ctrl.

The results are shown in Figure 20.2.

---

## Stata Results

```
. oneway b ctrl, tabulate

     Treated| Summary of Surv.TimeFrDate-of-Untreatabili
  w/Vit C? Or|
    Control?|       Mean    Std. Dev.        Freq.
 -----------+-----------------------------------
    Treatmt |     279.80       309.37           30
   Controls |      32.60        18.33           30
 -----------+-----------------------------------
      Total |     156.20       250.49           60

                    Analysis of Variance
    Source              SS         df       MS            F      Prob > F
 -----------------------------------------------------------------------
 Between groups     916617.60       1    916617.60      19.09     0.0001
 Within groups     2785330.00      58    48022.931
 -----------------------------------------------------------------------
    Total          3701947.60      59    62744.8746

 Bartlett's test for equal variances:   chi2(1) = 121.8158  Prob>chi2 = 0.000
```

**Figure 20.2** *One-way ANOVA, Effectiveness of Vitamin C Data*

ANOVA is based on the notion of the sums of the squares, which means the difference between each value and its mean, squared and totaled. This concept is the same one used to compute the variance and standard deviation. The results have several different sums of the squares, which are explained using the results in Figure 20.2.

- The *Between groups* sum of the squares (labeled SS, with a value of 916,617.60) is based on the deviation between each group's mean (we have two groups, the treatment group and the control group) and the overall mean for the entire data set. This difference is squared and totaled for each observation.

- The *Between groups* degrees of freedom (labeled df, with a value of 1) is K–1, where K is the number of categories in the independent variable. We have two categories (treatment and control); thus one degree of freedom.

- The *Between groups* mean square (labeled MS, with a value of 916,617.60 again) is the SS divided by the df.

- The *Within groups* sum of the squares (2,785,330.00) is the deviation of each observation from the mean for its category, again squared and totaled.

- The *Within groups* degrees of freedom is the number of observations minus the number of groups. Here it is 60 – 2 = 58.

- The *Within groups* mean square (48,022.931) is the sum of the squares divided by its degrees of freedom.

- The *F test* is the Mean Square (Between) divided by the Mean Square (Within). It has the same degrees of freedom as its numerator (the mean square/between) and denominator (the mean square/within). Here that would be 1 and 58 degrees of freedom. StataQuest has found that the probability that we could have received this result by chance from a population where there is no difference between the two groups is 0.0001— again 1 chance in 10,000.

## Bartlett's Test for Equal Variances

StataQuest also runs **Bartlett's test for equal variances**. This test has a null hypothesis that the variances of each category of the independent variable are equal and produces a chi-squared statistic, provided there are at least six observations per category, with degrees of freedom equal to one less than the number of categories. Where there are fewer than six observations per category, Bartlett's statistic has its own table. (See Kanji, 1993, 62–63.) The test assumes that the population is normally distributed. Bartlett's test for equal variances provides strong evidence that the variances are unequal, and we reject the null hypothesis. If you look at a box plot or dot plot by groups, with the group variable being *ctrl*, you can see that the data for those who fell into

---

## ■ TIP 22 ■ ■ ■ ■ ■ ■ ■ ■ ■ ■ ■ ■ ■ ■ ■ ■

### Additional Options (Command Mode Only)

The tabulate option will report means and standard deviation of the dependent variable for categorical breakdowns along with the ANOVA table of results. The Bonferroni, Scheffe, and Sidak multiple-comparison tests are also available as options. Examples:

```
. oneway b ctrl, tabulate
. oneway b ctrl, tabulate bonferroni
. oneway b ctrl, tabulate scheffe
. oneway b ctrl, tabulate sidak
```

---

the control group indicates on *b* that they, for the most part, died within a short time, while some of the treatment group members lived much longer. See Howell, 1992, 307–309 for suggestions when equal variances are untenable.

### Diagnostic Graphs

The ANOVA procedures also allow the user the option of running two diagnostic graphs. The **error-bar plot** provides the user with a graph of the mean of each group, with a bar the length of one standard error above and below the mean to indicate its dispersion. The horizontal line indicates the overall mean. The **graph of means** shows the group means along with the horizontal line, again indicating the overall mean. The graphs are useful in showing graphically how distinct the groups are.

## Two-Way ANOVA

Two-way analysis of variance compares the means of the categories of two independent variables with the overall mean for the entire data set. The assumptions are the same as with one-way.

The term **two-way** refers to the fact that we are comparing groups on two dimensions, each an independent variable. For example, let's assume we want to compare whether a subject was in the treatment or control group with the kind of cancer, either stomach or colon. StataQuest will produce a cross-tabulation showing the overall mean, the mean for each category of each variable, and the mean for each cell using the *Means and SDs by Group* and then *Two-way of means* procedure found under the *Summaries* menu. We will not duplicate those results here, but they indicate that the two kinds of cancer might be quite different. The colon cancer patients seem to live longer than the stomach cancer patients. Similarly, the treatment patients seem to live longer than the controls.

We can see whether these differences are statistically significant using the *Two-way ANOVA* procedure.

**1** From the *File* menu, choose *Open*.

**2** Enter the name of the file: `pauling`.

**3** From the *Statistics* menu, choose *ANOVA* and then *Two-way*.

**4** **Windows:** For the dependent variable, click on *b*. The two categorical variables should be *ctrl* and *kind*. Click on OK. (The Windows default is to add an interaction term.)

**DOS:** In answer to the questions, the dependent variable should be b, and the two categorical variables should be `ctrl` and `kind`. (The DOS default is **not** to add an interaction term.)

The results are shown in Figure 20.3.

## Stata Results

```
. anova b kind ctrl kind*ctrl

                  Number of obs =        60     R-squared     =  0.3027
                  Root MSE      = 214.704     Adj R-squared =  0.2653

       Source |   Partial SS    df       MS               F      Prob > F
    ----------+----------------------------------------------------------
        Model |   1120463.66     3   373487.888            8.10    0.0001
              |
         kind |   103220.514     1   103220.514            2.24    0.1402
         ctrl |   821846.883     1   821846.883           17.83    0.0001
    kind*ctrl |   100625.549     1   100625.549            2.18    0.1452
              |
     Residual |   2581483.94    56   46097.9274
    ----------+----------------------------------------------------------
        Total |   3701947.60    59   62744.8746
```

**Figure 20.3** *Two-way ANOVA, Effectiveness of Vitamin C Data*

Figure 20.3 indicates that we can reject the null hypothesis that there is no difference between the treatment and the control groups with a high degree of confidence, even when controlling for the kind of cancer.

## Diagnostic Graphs

After the two-way ANOVA procedure, a pop-up box (in SQ for Windows only) asks whether you want to run any of four different diagnostic plots:

- The **interaction plot** is a plot with one of the categorical variables on the X-axis, the dependent variable measured on the Y-axis, and separate lines for each category of the other categorical variable. These lines are a graphical depiction of the cell means. The cell means can also be obtained by choosing the *Means and SDs by Group* and then *Two-way of means* procedure from the *Summaries* menu.

- An **error bar plot for one categorical variable**—this graph is the same as that reported for one-way ANOVA.

- An **error bar plot for the other categorical variable**—this graph is the same as that reported for one-way ANOVA..

- A **graph of means for the two categorical variables** computed separately on the same graph—this graph is the same as that reported for one-way ANOVA.

For more information on the the effects of vitamin C on survival times for patients with cancer, consult Cameron and Pauling (1976, 1978) or examine and analyze the much more extensive data set reprinted in Andrews and Herzberg (1985, 203–207).

# Repeated Measures ANOVA

Repeated measures ANOVA is a method that enables the data analyst to take account of the fact that some experiments necessitate multiple measurements on each subject. The migraine headache experiment analyzed, for example, has nine subjects, each of whom was a migraine-headache sufferer. Each subject participated in a five-week experiment, with five measurements of the frequency and intensity of migraines collected. The task of repeated measures ANOVA is to separate the differences between the subjects from the differences over the five weeks.

The migraine data is from Howell (1992, 437), who reports an experiment on minimizing the frequency and duration of migraine headaches through relaxation techniques. Each of nine subjects participated in several weeks of training and then a five-week period of measurements. The first two measurements are a baseline, and the last three are the effects of the relaxation training. Thus, we would be looking for a trend or effect in the variable for the week of the experiment (called week in *migraine.dta*) while controlling for the effects and differences among the individual subjects (the variable subject).

**Table 20.1** *Repeated Measures ANOVA Data, Migraine Data Set in Original Form*

| Subject | Week 1 | Week 2 | Week 3 | Week 4 | Week 5 |
|---------|--------|--------|--------|--------|--------|
| 1 | 21 | 22 | 8 | 6 | 6 |
| 2 | 20 | 19 | 10 | 4 | 4 |
| 3 | 17 | 15 | 5 | 4 | 5 |
| 4 | 25 | 30 | 13 | 12 | 17 |
| 5 | 30 | 27 | 13 | 8 | 6 |
| 6 | 19 | 27 | 8 | 7 | 4 |
| 7 | 26 | 16 | 5 | 2 | 5 |
| 8 | 17 | 18 | 8 | 1 | 5 |
| 9 | 26 | 24 | 14 | 8 | 9 |

**Table 20.2** *First Five Observations, Data as Entered*

| | first 5 observations | |
|-----------|------|--------------|
| Subject # | Week | Migraine Data |
| 1 | 1 | 21 |
| 1 | 2 | 22 |
| 1 | 3 | 8 |
| 1 | 4 | 6 |
| 1 | 5 | 6 |

**Table 20.3** *Last Five Observations, Data as Entered*

| | last 5 observations | |
|-----------|------|--------------|
| Subject # | Week | Migraine Data |
| 9 | 1 | 26 |
| 9 | 2 | 24 |
| 9 | 3 | 14 |
| 9 | 4 | 8 |
| 9 | 5 | 9 |

StataQuest requires that the data **not** be arranged in what might seem to be the logical method, that is, like that in Table 20.1, but instead to be formatted so that each observation for each week is a single line of data. This arrangement necessitates that you have a variable for subject number, a variable for the week, and a variable for the actual data. Tables 20.2 and 20.3 show the first five and last five observations of the data.

This format is required for repeated measures ANOVA and is the way that the *migraine.dta* data set is formatted, with five observations of nine subjects. *migraine.dta* thus has 45 total observations and variables for both subject number and week of the experiment.

To run repeated measures ANOVA,

1. Ensure that your data follows a format that SQ can analyze for repeated measures analysis. See the discussion immediately preceding.

2. From the *File* menu, choose *Open*.

3. Enter the name of the file: migraine.

4. From the *Statistics* menu, choose *ANOVA* and then *Repeated measures*.

5. **Windows:** For the dependent variable, click on *migraine*. The subject ID variable is *subject*; the categorical variable is *week*. Click on OK.

   **DOS:** In answer to the questions, the dependent variable should be migraine, the subject ID variable is subject; the categorical variable is week.

The results are shown in Figure 20.4.

**Stata Results**

```
. anova migraine subject week
                   Number of obs =       45    R-squared     =  0.9272
                   Root MSE      = 2.68328    Adj R-squared =  0.8999

      Source |  Partial SS    df      MS             F     Prob > F
    ---------+----------------------------------------------------
       Model |  2935.91111    12   244.659259       33.98    0.0000
             |
     subject |  486.711111     8  60.8388889        8.45    0.0000
        week |     2449.20     4      612.30        85.04    0.0000
             |
    Residual |      230.40    32        7.20
    ---------+----------------------------------------------------
       Total |  3166.31111    44  71.9616162
```

**Figure 20.4** *Repeated Measures ANOVA, Migraine Headache Experiment*

Figure 20.4 indicates that the effect over time for the *week* variable is highly significant once we control for the individual subjects. The model as a whole is highly significant, and the R-squared from the regression explains about 90% of the variance.

---

### ■ TIP 23 ■ ■ ■ ■ ■ ■ ■ ■ ■ ■ ■ ■ ■ ■ ■

## Additional Options (Command Mode Only)

Virtually all ANOVA commands in StataQuest can be replicated by using the regression procedures. You define categorical independent variables as so-called dummy variables (0,1 codes) and entering them into the equation. That topic is beyond the scope of this book, but ANOVA has a *regress* option that allows you to see the same results in the language and format of regression output. You can obtain the results from the *regress* option in *either* of two ways:

■ Run the analysis of variance from within the menu structure, go to command mode or the Command window, and type the command `regress`. `regress` will output on the screen the results of the last regression, which in this case is your ANOVA, in regression format.

■ Run the entire analysis of variance in command mode, using the ANOVA command. Enter the following command in command mode or a Command window: `anova a kind ctrl , regress`

---

## N-Factor ANOVA & ANOCOVA

The option for *N-Factor ANOVA & ANOCOVA* on the *ANOVA* submenu allows the user to add additional interactions, including those at higher levels than the two-level interaction just considered, and variables that are continuous, allowing the user to run analysis of covariance (ANOCOVA) models.

We shall run a model based on the data from the low birth weight study (*lowbw.dta*). Our dependent variable is the weight of the baby in grams (*bwt*). We have three categorical variables: whether the mother smoked during pregnancy (*smoke*), whether the mother has a history of hypertension (*ht*), and whether the mother has a history of uterine irritability (*ui*). In addition, we have two continuous covariates, the weight of the mother at the last menstrual period (*lwt*), and the age of the mother (*age*).

1  From the *File* menu, choose *Open*.

2  Enter the name of the file: `lowbw`.

3  From the *Statistics* menu, choose *ANOVA* and then *N-factor ANOVA & ANOCOVA*.

4  **Windows:** For the dependent variable, click on *bwt*. The three categorical variables should be *smoke*, *ht*, and *ui*. The two continuous variables should be *lwt* and *age*.

**Build Interaction Terms:** In this area, you can build interactions that go beyond the two-level interactions that SQ automatically builds and tests in the Two-way ANOVA procedure.

First, we will **add the two-way interactions** by clicking on each pair of two variables and then clicking on "Add term." At that point the interaction should appear in the box at the bottom .

Then add any **three or higher level interactions**. We will build a three level interaction among *smoke, ht,* and *ui.* Click on each of these in the box to the right of "Build Interaction Terms:" and then click on the "Add term" box. Notice that "smoke*ht*ui" appears in the box next to "Interactions." You can build more of these terms if we were to include more categorical variables. Click on OK to run the procedure.

**DOS:** In answer to the questions, the dependent variable should be *bwt,* and everything else is a "term." The three categorical variables should be smoke, ht, and ui, and the two continuous variables should be lwt and age. Type in the two-way interactions (smoke*ht, ht*ui, smoke*ui). Then type in the three-way interaction: smoke*ht*ui.

The results are shown in Figure 20.5.

## Stata Results

```
. anova bwt    smoke ht ui    lwt age   smoke*ht*ui smoke*ht smoke*ui
ht*ui, cat(>   smoke ht ui)
               Number of obs =     189    R-squared      =   0.1840
               Root MSE      = 671.167    Adj R-squared =   0.1524
```

| Source | Partial SS | df | MS | F | Prob > F |
|---|---|---|---|---|---|
| Model | 18382958.9 | 7 | 2626136.99 | 5.83 | 0.0000 |
| | | | | | |
| smoke | 76735.9532 | 1 | 76735.9532 | 0.17 | 0.6803 |
| ht | 3990236.53 | 1 | 3990236.53 | 8.86 | 0.0033 |
| ui | 6164842.93 | 1 | 6164842.93 | 13.69 | 0.0003 |
| lwt | 3145561.02 | 1 | 3145561.02 | 6.98 | 0.0090 |
| age | 125899.087 | 1 | 125899.087 | 0.28 | 0.5977 |
| smoke*ht*ui | 0.00 | 0 | | | |
| smoke*ht | 82394.5596 | 1 | 82394.5596 | 0.18 | 0.6694 |
| smoke*ui | 947304.207 | 1 | 947304.207 | 2.10 | 0.1487 |
| ht*ui | 0.00 | 0 | | | |
| | | | | | |
| Residual | 81534093.7 | 181 | 450464.606 | | |
| Total | 99917052.6 | 188 | 531473.684 | | |

**Figure 20.5** *N-Factor ANOVA and ANOCOVA, Low Birth Weight Study Data*

The results in Figure 20.5 indicate that the variables *ht*, *ui*, and *lwt* are all significant when we control for *smoke* and the appropriate interactions; *age* and *smoke* are not significant. Two of the two-way interactions are not significant, and the other two turn out to be perfectly correlated with variables already in the model and are thus listed with zero for both the Partial Sum of the Squares ("Partial SS") and the degrees of freedom ("df"). SQ automatically takes care of such problems, dropping the variables not needed and running the rest of the model.

## ▓ EXERCISES

20.1   Use the data set *caffeine.dta* to run a one-way ANOVA on the number of finger taps per minute by the amount of caffeine the subjects had been given. Do the results make sense? Run the appropriate follow-up graphs and interpret. Be sure to use *Describe variables* and *Dataset information* from the *Summaries* menu to ensure that you understand the data.

20.2   Use the *deaths.dta* data set, in which each observation is a famous man and the number of months he died before or after his birthday. Use the *deaths* variable and the *group* variable on both a One-way ANOVA and a two-sample t-test to collect evidence on the question of whether famous men die just before or just after their birthdays. Why do you suppose this phenomenon occurs?

20.3   Use the data set *pres.dta* to examine the question of whether U.S. Presidents, Popes, or British monarchs live longest after inauguration.

20.4   Use the data set *lowbw.dta* to run a two-way ANOVA of your choice.

20.5   Use the data set *lowbw.dta* to run a N-Factor ANOVA and ANOCOVA of your choice.

# *StataQuest's Functions*

## Mathematical Functions

**abs(x)**  Absolute value. The absolute value of a number is the number with a positive sign. If we had a variable called *feeling1* which ran from -7 (extremely negative feeling) to zero (neutral feeling) to +7 (extremely positive feeling), and we wanted a measure of how strongly the respondent felt about the issue, without regard to whether he or she felt positively or negatively, we might use *abs(feeling1)* to give us the absolute value of the feeling, disregarding the negative signs for those respondents who were negative.

**exp(x)**  exp(x) yields e to the power of x. Thus,

$exp(2) = e^2 = 2.7182818$ squared $= 7.3890559$

$exp(3) = e^3 = 2.7182818$ cubed $= 20.085536$

**ln(x)**  Natural logarithm, or log to the base e. To obtain logs to the base 10, use the following formula: *ln(x) / ln(10)*.

**mod(x,y)**  Modulus of x with respect to y. The modulus is the remainder of x when divided by y. Thus, mod(10,3) = 1.

*sqrt(x)*        Square root. The *sqrt* function is useful in transforming skewed
                variables. It, like the *log* function, tends to reduce the incidence
                of data at the extremes.

**atan(x)**      arc-tangent returning radians

**cos(x)**       cosine of radians

**gammap(a,x)**  incomplete gamma P(a,x)

**ibeta(a,b,x)** incomplete beta $I_x(a,b)$

**lngamma(x)**   ln(Gamma(x))

**log(x)**       natural logarithm, same as ln(x)

**sin(x)**       sine of radians

**cos(x)**       cosine of radians

# Statistical Functions

**uniform()** To generate random data over the open interval 0 to 1, use the
*uniform()* function. You must include the parentheses although nothing
goes in them. The *uniform()* function uses the same starting number
and sequence each time StataQuest is started. The starting value can be
changed with the command mode command: set seed X, where X
is a large odd, positive integer. Note: SQ for Windows has this function incorporated into the *Data/Generate/Replace/Random/Uniform*
command.

**invnorm(uniform())** To generate normally distributed random numbers with
a mean of 0 and standard deviation of 1, use the formula *invnorm
(uniform())*. Note: SQ for Windows has this function incorporated
into the *Data/Generate/Replace/Random/Normal* command.

**M + S * invnorm(uniform())** To generate normally distributed random numbers
with a mean of M and standard deviation of S, use the formula *M + S *
invnorm(uniform())*, where you insert your proposed mean for M and
your proposed standard deviation for S. If the mean is to be 10 and the
standard deviation 2, then the formula would be *10 + 2 * invnorm
(uniform())*. Again, this function is incorporated in SQ for Windows
into the *Data/Generate/Replace/Random/Normal* command.

**binomial(n,k,pi)** Probability of observing k or more successes in n trials when
the probability of a single success is pi.

**chiprob(df,x)** Cumulative chi-squared statistic with df degrees of freedom and
value x

**fprob(df1,df2,f)** Cumulative F distribution with df1 numerator and df2
denominator degrees of freedom

**invbinomial(n,k,p)** Inverse binomial; for p > .5, returns probability pi such that the probability of observing k or more successes in n trials is p; for p < .5, returns probability pi such that the probability of observing k or fewer successes in n trials is 1-p.

**invt(df,p)** Inverse two-tailed cumulative t-distribution.

**normprob(z)** Cumulative normal distribution.

**tprob(df,t)** Student's two-tailed t-distribution with df degrees of freedom; returns probability $|T| > |t|$.

# String Functions

| | |
|---|---|
| **index(s1,s2)** | returns the position in s1 in which s2 is first found or zero if s1 does not contain s2 |
| **length(s)** | returns the actual length of the string |
| **lower(s)** | returns the lowercased variant of s |
| **ltrim(s)** | returns s with any leading blanks removed |
| **real(s)** | converts s into a numeric argument |
| **rtrim(s)** | returns s with any trailing blanks removed |
| **string(n)** | converts n into a string |
| **substr(s,n1,n2)** | returns the substring of s starting at the n1th column for a length of n2 |
| **trim(s)** | returns s with any leading and trailing blanks removed |
| **upper(s)** | returns the uppercased variant of s |

# Special Functions

**group(x)** Creates a categorical variable that divides the data into x as nearly equally sized subsamples as possible. If you want to divide a variable into three equally sized subgroups, give *group(3)* when StataQuest asks for its formula.

**int(x)** Yields the integer obtained by truncating x. Truncating means disregarding whatever is to the right of the decimal point and is different from rounding up or down. 10.8 would be truncated to 10, whereas it would be rounded to 11. The number 20.2 would be both truncated and rounded to 20.

**max(x1, x2, ... xn)** Yields the maximum of any of the variables x1, x2, and so on. Suppose you have four indicators of an underlying concept the respondent could practice very infrequently to very frequently. You want to know what the most frequent response your respondent had to any of the four variables. You could use the *max* function to ascertain that information. Suppose the variables x1, x2, x3, and x4 are all coded from 1 (infrequent) to 5 (very frequent). Then *max(x1, x2, x3, x4)* would yield the maximum score on any of the four variables. We would, of course, put that score in a column separate from the four original variables so as not to wipe any of them out in the process.

**min(x1, x2, ... xn)** Yields the minimum of any of the variables included within the parentheses. (See example under *max.*)

**round(x,y)** Yields x rounded to units of y. If variable x1 is equal to 4.8 in a particular case and we use the function *round(x1,1)*, we will get 5 for that particular case. *round(x1,.01)* will round x1 to the nearest hundredth.

**[ _n–x]** To obtain a lagged variable, you write a formula with [ _n-x], where x is the number of lagged periods. To lag the variable gnp by one period, enter the formula gnp[ _n-1]; for three periods, gnp[ _n-3].

**autocode(x,ng,xmin,xmax)** partitions the interval from xmin to xmax into ng equal length intervals and returns the upper bound of the interval that contains x

**cond(x,a,b)** returns a if x evaluates to true (not 0) and b if x evaluates to false (0)

**float(x)** returns the value of x rounded to float

**sign(x)** returns missing if x = . , -1 if x < 0, 0 if x = 0, and 1 if x > 0

**sum(x)** returns the running sum of x, treating missing values as zero

For more information, consult Hamilton (1993) or the *Stata Reference Manuals* (1995). StataQuest also contains extensions to the generate command in command mode; these provide many additional functions useful for those doing advanced data analysis.

# ▪ ▪ ▪ ▪ ▪ ▪ ▪ BIBLIOGRAPHY

Allswang, John M. 1991. *California Initiatives and Referendums, 1912–1990.* Los Angeles, California: Edmond G. "Pat" Brown Institute of Public Affairs.

Andrews, D. F., and A. M. Herzberg. 1985. *Data, A Collection of Problems from Many Fields for the Student and Research Worker.* New York: Springer-Verlag.

Anscombe, F. 1973. "Graphs in Statistical Analysis," *American Statistician,* February 27, 17–21.

Barone, Michael, and Grant Ujifusa. 1993. *The Almanac of American Politics 1994.* Washington, D.C.: *National Journal.*

Bollen, Kenneth, and Robert Jackman. 1990. "Regression Diagnostics: An Expository Treatment of Outliers and Influential Cases." In *Modern Methods of Data Analysis.* Edited by John Fox and J. Scott Long. Newbury Park, California: Sage.

Bowen, Bruce and Herbert Weisberg. 1980. *An Introduction to Data Analysis.* San Francisco: W.H. Freeman.

Cameron, E., and L. Pauling. 1976. "Supplemental Ascorbate in the Supportive Treatment of Cancer: Prolongation of Survival Times in Terminal Human Cancer." *Proc. Natl. Acad. Sci. U.S.A.* 73:3685–3689.

Cameron, E., and L. Pauling. 1978. "Supplemental Ascorbate in the Supportive Treatment of Cancer: Revaluation of Prolongation of Survival Times in Terminal Human Cancer." *Proc. Natl. Acad. Sci. U.S.A.* 75:4538–4542.

Chambers, John M., et al. 1983. *Graphical Methods for Data Analysis.* Belmont, California: Wadsworth.

Cleveland, William S. 1985. *The Elements of Graphing Data.* Monterey, California: Wadsworth Advanced Books and Software.

Conover, W. J. 1980. *Practical Nonparametric Statistics.* New York: John Wiley.

D'Agostino, R. B., et al. 1990. "A Suggestion for Using Powerful and Informative Tests of Normality." *The American Statistician* 44:316–321.

D'Agostino, R. B., et al. 1991. "Comment on Tests of Normality." *Stata Technical Bulletin* 3:20.

Daniel, Wayne W. 1978. *Applied Nonparametric Statistics.* Boston: Houghton Mifflin.

Devore, Jay, and Roxy Peck. 1993. *Statistics: The Exploration and Analysis of Data.* Belmont, California: Duxbury Press.

Dye, Thomas. 1990. *American Federalism: Competition Among Governments.* Lexington: Heath Lexington, 36.

Farnum, Nicholas R. 1994. *Modern Statistical Quality Control and Improvement.* Belmont, California: Duxbury.

Fienberg, Stephen E. 1973. "Randomization for the Selective Service Draft Lotteries." In *Statistics by Example: Finding Models*. Edited by F. Mosteller, W. Kruskal, R. Link, R. Pieters, and G. Rising. Menlo Park, California: Addison-Wesley.

Fox, John. 1991. *Regression Diagnostics*. Newbury Park, California: Sage.

Gould, W. W. 1991. "Skewness and Kurtosis Tests of Normality." *Stata Technical Bulletin* 1:20–21.

Gould, W. W. 1992. "Final Summary of Tests of Normality." *Stata Technical Bulletin* January, 5:10–11.

Gould, W. W., and W. Rogers. 1991. "Summary of Tests of Normality." *Stata Technical Bulletin* 3:20–23.

Hamilton, Lawrence C. 1990. *Modern Data Analysis: A First Course in Applied Statistics*. Belmont, California: Duxbury.

Hamilton, Lawrence C. 1992. *Regression with Graphics, A Second Course in Applied Statistics*. Belmont, California: Duxbury.

Hamilton, Lawrence C. 1993. *Statistics With Stata 3*. Belmont, California: Duxbury.

Hand, D. J., et. al. 1994. *A Handbook of Small Data Sets*. London: Chapman and Hall.

Hanushek, Eric A. and John Jackson. 1977. *Statistical Methods for Social Scientists*. New York: Academic Press.

Hays, William H., and Robert L. Winkler. 1971. *Statistics: Probability, Inference and Decision*. New York: Holt, Rinehart and Winston.

Hoaglin, David C., Frederick Mosteller, and John W. Tukey. 1983. *Understanding Robust and Exploratory Data Analysis*. New York: John Wiley.

Hosmer, David W., Jr. 1989. *Applied Logistic Regression*. New York: John Wiley.

Hosmer, David W., and Stanley Lemeshow, 1989. *Applied Logistic Regression*. New York: John Wiley.

Howell, David C. 1992. *Statistical Methods for Psychology*. 3rd ed. Belmont, California: Duxbury.

Johnson, Robert. 1992. *Elementary Statistics*, 6th ed. Boston: PWS-Kent.

Judge, G. G., et al. 1985. *Theory and Practice of Econometrics*. 2nd ed. New York: John Wiley.

Kanji, Gopal K. 1993. *100 Statistical Tests*. Newbury Park, California: Sage.

Labich, Kenneth. 1993. "The Best Cities for Knowledge Workers." *Fortune*. November 15, 50–78.

Lebergott, Stanley. 1993. *Pursuing Happiness: American Consumers in the Twentieth Century*. Princeton, New Jersey: Princeton University Press.

Liao, Tim Futing. 1994. *Interpreting Probability Models: Logit, Probit, and Other Generalized Linear Models*. Thousand Oaks, California: Sage.

Mosteller, Frederick and Robert E. K. Rourke. 1973. *Sturdy Statistics, Non-parametrics and Order Statistics*. Reading, Massachusetts: Addison-Wesley.

*The New Columbia Encyclopedia.* 1975. New York: Columbia University Press.

Royston, J. P. 1991a. "Tests for Departure from Normality." *Stata Technical Bulletin* 2:16–17.

Royston, J. P. 1991b. "Shapiro-Wilk and Shapiro-Francia Tests." *Stata Technical Bulletin* 3:19.

Royston, J. P. 1991c. "Comment on sg3.4 and an Improved D'Agostino Test." *Stata Technical Bulletin* 3:23–24.

Royston, J. P. 1991d. "A Response to sg3.3: Comment on Tests of Normality." *Stata Technical Bulletin* 4:8–9.

Siegel, Sidney. 1956. *Nonparametric Statistics for the Behavioral Sciences.* New York: McGraw-Hill.

Smith, Edward S. 1947. *Control Charts: An Introduction to Statistical Quality Control.* New York: McGraw-Hill.

SRM. 1993. *Stata Reference Manual.* Stata Corporation. *Stata Release 3.1, Reference Manual. Volumes I-III.* College Station, Texas: Stata Corporation, 702 University Drive East.

*Stata Reference Manuals.* 1995. Stata Corporation. *Reference Manual, Release 4, Stata.* College Station, Texas: Stata Corporation, 702 University Drive East.

Tukey, John W. 1977. *Exploratory Data Analysis.* Menlo Park, California: Addison-Wesley.

United Nations Development Programme (UNDP). 1990. *Human Development Report 1990.* New York: Oxford University Press.

U.S., Bureau of the Census. 1994. *Statistical Abstract of the United States: 1994.* Washington, D.C.: Government Printing Office.

Williams, Bill. 1987. *A Sampler of Sampling.* New York: John Wiley.

World Resources Institute. 1993. *The 1993 Information Please Environmental Almanac.* New York: Houghton Mifflin.

Wright, John W. 1989. *The Universal Almanac 1990.* New York: Andrews and McMeel.

Wright, John W. 1992. *The Universal Almanac 1993.* Kansas City, Missouri: Universal Press.

# INDEX ■ ■ ■ ■ ■ ■ ■ ■ ■ ■

0.0000, interpretation of, 74
3.29e–06, interpretation of, 218

**A**
Added variable plots, 249
Adding observations, 39–42
Adding variables, 39–42
Additional options, R charts, X-bar charts, 171
Adjacent values, in box plots, 112–113
Adjusted R-square, regression, 238
Alphanumeric variable, 2–3, 48, 59
Alphanumeric variables, converting, 48
ANOCOVA, 290–292
ANOVA, 280–292
ANOVA, n-factor, 290–292
ANOVA, one-way, 282–284
ANOVA, repeated measures, 287–289
ANOVA, two-way, 285–287
ASCII file, 16–17
ASCII files, inputting, 32–33
Autocorrelation, 265
AV plots, 249, 260–263

**B**
Backward elimination, stepwise regression, 257–260
Bar chart comparisons, 126–128
Bar chart comparisons by group, 129–130
Bar charts, 123–133
Bar charts by group, 124–126
Bartlett's test for equal variances, 284
Basic statistics, 60–67
Bibliography, 297–299

Binomial confidence interval, calculator, 217–218
Binomial test, 184–187
Binomial test, normal approximation, 186–187
Bins, in histograms, 82–84
Box and one-way plot, 111, 114–115
Box plot and one-way plot comparisons, 119–121
Box plot by group, 115–117
Box plot comparisons, 117–119
Box plots, 111–114
Buttons, listed, 12–13

**C**
C charts, 162–163
Calculator, regular, 222–223
Calculator, RPN, 223
Calculator, statistical, 204–225
Case, 2
Cases, adding, 40–42
Cases, deleting, 42–43
Categorical variables, 70
Cells, changing contents of, 37–40
Changing cell contents, 37–40
Changing groups of cells, 38–40
Checking a cross-tab, 224
Chi-squared distribution, from a cross-tab in a book, 224
Chi-squared interpretation, in cross-tabulation, 74, 78
Close, button, 25
Column percentages, 72, 74
Command mode, 5, 41, 147
Command mode, ANOVA, 290
Compatibility of Stata, SQ files, 30

Confidence interval, post-regression diagnostics, 242–243
Confidence intervals (mean), 65–67
Confidence intervals, calculator, 215–219
Confidence intervals, regression, 240
Contents of data sets, 55–60
Continuous variable histograms, 81–88
Continuous variable histograms, by group, 89–91
Control (C) charts, 162–163
Correlation matrix, 229
Correlation, 226–234
Cramer's V, 74
Creating random variables, 33–35
Cross-tab, checking using the statistical calculator, 224
Cross-tabulation, 72–78
Cross-tabulation by a third variable, 76–78

**D**
Data elements, 2–3
Data files, 16–17
Data set information, 56–58
Data set labels, 31–32
Data sets, contents, 55–59
Data, entering, 21–27
Default fonts, 7, 8, 9
Delete, button, 25
Deleting observations, 42–43
Deleting variables, 42–43
Describe variables, 55–56
Diagnostics, regression, 236, 241–250
Dialog box, regression diagnostics, 243
Dialog boxes, 10–11

Dichotomous variable, 36
Discrete variable histograms, 88
Discrete variable histograms by group, 91
DOS, basic rules, 13
DOS, installation, 4, 5
Dot plots, 105–108
Durbin-Watson statistic, 265

**E**

edit command, 39, 40
Editor, 21–27
Editor, buttons, 24–25
Editor, changing cells, 37–40
Editor, close, 25, 34
Editor, delete, 25
Editor, DOS, 26–27
Editor, hide, 25
Editor, sort, 25
Entering data, 21–27
Equality of populations test, 197–198, 201
Equals sign (double), 39
Error-bar plot, ANOVA, 285, 287
Exiting from SQ, 4, 5
Exiting the editor, 25,34
External files, inputting, 27–28

**F**

F-statistic, regression, 238
Fences, box plot, 112–113
File compatibility, 30
File maintenance, 17
File size, 2
Files, different, 16–17
*fit* command, 249, 260–261
Fonts, 7, 8, 9
Formulas, 43–46
Forward inclusion, stepwise regression, 257–258
Fraction defective (P) charts, 164
Frequency distributions, 70–72
Functions, 43, 46, 293–296

**G**

Gamma, 74–75

Generate/Replace, 43–47
Graph files, 15–17
Graph of means, ANOVA, 285, 287
Graphs, placing in word processor, 14
Group, bar charts by, 124–126
Group, box plots by, 115–117
Group, dot plots by, 108
Group, histograms by, 89–91
Group, means and standard deviations, 61–64

**H**

Help, 13
Hide, button, 25
Histograms, 81–91

**I**

"if" statements, 39
"in" statements, 40
Individual Y values, predicting, regression, 243
Influence statistics, multiple regression, 263–264
Information, data sets, 55–58
Inputting external files, 27–28
Inputting SAS files, 31
Inputting SPSS files, 31
Inputting Stata files, 30
Inputting statistical package files, 31
Installation of SQ, 3–4
Inter-Quartile Range, 112–113
Interaction plot, ANOVA, 287
Inverse statistical tables, 221–222
IQR, 112–113

**J**

Jittering, scatter plots, 116

**K**

Kendall's tau-b, 75
Kolmogorov-Smirnov tests, 199–201
Kruskal-Wallis equality of populations test, 197–198, 201, 280–282
Kurtosis coefficient, 64

**L**

Labeled to string, 47
Labeling data, 31–33
Lagging variables, 46
LCL lines, 166
Leverage plots, 261–262
Likelihood-ratio chi-squared, 74
Limits of SQ, 1
Linear relationship, correlation, 230
Linear relationships, regression, 235
Listing data, 58–60
Log files, 14–15
Logistic regression, 279
Logit, 270
Logs, 10, 44, 65, Appendix A

**M**

Maintenance of files, 17
Mann-Whitney test, 196–197, 201
Maximum file size, 2
Mean square error, regression, 239
Mean value of Y, predicting, regression, 243
Means, 60–64
Means and standard deviations, by group, 61–64
Means, one-way, 61
Means, two-way, 62
Measurement variables, 69, 71
Median and Percentiles, 64–65
Memory, data in, 22
Menu items, listed, 12
Menu mode 5–13
Model sum of the squares, regression, 238
Modifying data, 36–53
Multiple regression, 252–267

**N**

N-factor ANOVA, 290–292
Names, variable, 33
Nonparametric ANOVA, 280–282

Nonparametric tests, 172, 190–203

Normal approximation to binomial test, 186–187

Normal approximation, one-sample test of proportion, calculator, 212

Normal curve overlay, in histograms, 84–85

Normal quantile plot, regression residuals, 246–247

Normal quantile plots, 92–94

Normality, testing for, 182–183

Notes on data sets, 56–58

Numbers, 48

Numbers, scientific, 218

**O**

Observation, 2

Observations, adding, 40–42

Observations, deleting, 42–43

Odds, logistic regression, 270

Odds-ratios, logistic regression, 271

One-sample and two-sample tests of variance, 180–181

One-sample Kolmogorov-Smirnov test, 199, 201

One-sample normal test, calculator, 204–206

One-sample t-test, 172–177

One-sample t-test, calculator, 207–208

One-sample test of proportion, 184–185

One-sample test of proportion, calculator, 210–212

One-sample test of proportion, normal approximation, calculator, 212

One-sample test of variance, 180–181

One-sample test of variance, calculator, 215

One-way ANOVA, 282–284

One-way means, 61

One-way plots, 111, 116

Order of variables, 158

Outliers, 112–113, 116

Overflows, stem-and-leaf plots, 105

**P**

P charts, 164, 166

Paired t-test, 180

Parametric tests, 172–189

Partial regression coefficients, 254

Partial regression leverage plots, 261–262

Pearson correlation, 227–230

Percentage difference interpretation, 74, 76

Percentages, column, 72, 74

Percentages, row, 72, 74

Percentiles, 64

Plane, regression, 256

Plot fitted model, regression, 242

Plot residual versus an X, regression, 247

Plot residual versus prediction, regression, 245

Plot, regression, 242

Plots, added variable, regression, 249

Plotting symbols, 147

Poisson confidence interval, calculator, 219

Post-regression diagnostics, 236, 241–250

Predicted values, regression, 240

Predicted values, saving, regression, 248–249

Prediction interval, post-regression diagnostics, 242–243

Preserve (in editor), 24–25

Printing graphs, 15–16

Printing results, 14–15

Probability of 0.0000, interpretation, 74

Pruning stem-and-leaf plots, 101–102

**Q**

Quality control charts, 161–171

**R**

R charts, 167–168

R-square, regression, 238

Random variables, 28–30

Range (R) charts, 167–168

Range, from/to, in histograms, 92

Rank correlation, 230–233

Recoding data, 47–50

Regression diagnostics, 236, 241–250, 253–257

Regression plane, 256

Regression, logistic, 279

Regression, multiple, 252–267

Regression, robust, 264–265

Regression, simple, 235–251

Regular calculator, 222–223

Regular correlation, 227–230

Renaming variables, 33

Repeated measures ANOVA, 287–289

Requirements, 1

Reserved words, 33

Residual sum of the squares, regression, 238

Residual versus an X, plot, regression, 247

Residual versus prediction, plot, regression, 245

Residuals, normal quantile plot of, 246–247

Residuals, regression, 245–246

Residuals, saving, 248–249

Restore (in editor), 25

Results window, 5–8

Results, placing in word processor, 14

Reverse Polish Notation, 223

Robust regression, 264–265

Root MSE, regression, 239

Row percentages, 72, 74

RPN calculator, 223

**S**

SAS files, inputting, 31

Saving and printing stem-and-leaf plots and dot plots, 108

Saving graphs, 16

Scaling option, time series plots, 154

Scatter plot matrix, 145–147, 148

Scatter plots, 134–149
Scatter plots, by group, 143
Scatter plots, naming points, 138–140
Scatter plots, symbols scaled to a third variable, 142–143
Scatter plots, with regression lines, 140–141
Scientific notation, 218
Session on SQ, structure, 18
Setting text size, graphs, 148
Shapiro-Francia W' test, 182–183
Shapiro-Wilk W test, 182–183
Sign test, 190–193, 201
Signed-Ranks test, 194–195, 201
Simple regression, 235–251
Single-sample Kolmogorov-Smirnov test, 199, 201
Skewness coefficient, 64
Slope coefficient, regression, 240
Sorting the data, correlation, 233
Sorting, 25
Spearman correlation, 230–233
SPSS files, inputting, 31
Stacked bar charts, 131
Standard calculator, 222–223
Standard deviation, 60–64
Standard error of the constant, regression, 240
Standard error of the slope, regression, 240
Starting SQ, 4
Stata, 18–19
Stata file compatibility, 30
Stata files, inputting, 30
StataQuest file compatibility, 30
Statistical calculator, 204–225
Statistical tables, 219–221
Statistical tables, inverse, 221–222
Stem-and-leaf plots, 97–105, 108–109
Stem-and-leaf plots, overflows, 105

Stems, stem-and-leaf plots, 99
Stepwise regression, 257–260
Stopping SQ, 4–5
String to labeled, 47
String variable, 2–3, 59
Subsetting observations, 50–52, 174–177
Symbols, plotting, 147, 148

**T**
Tables, 69–79
Tables, statistical, 219–221
Tables, statistical, inverse, 221–222
Testing for normality, 182–183
Tests of variance, calculator, 215
Text files, inputting, 27–28
Text size, graphs, 148
Time series plots, 150–160
Time series, regression, 265
Total sum of the squares, 238
Trends, graphing, 150
TSS, regression, 238
Two-sample Equality of Distributions test, 199–201
Two-sample Kolmogorov-Smirnov test, 199–201
Two-sample normal test, 206–207
Two-sample t-test, 178–180
Two-sample t-test, calculator, 208–210
Two-sample test of proportion, calculator, 213–215
Two-sample test of proportions, 187–188
Two-sample test of variance, 180–181
Two-sample test of variance, calculator, 215
Two-way ANOVA, 285–287
Two-way means, 62

**U**
UCL lines, 166
Univariate statistics, 60–65
Utilities, 17

**V**
Value labels, 31
Variable labels, 31
Variable names, 33
Variable order, 158
Variable, alphanumeric, 2–3, 59
Variable, defined, 2–3
Variable, string, 2–3, 59
Variables, adding, 40–42
Variables, categorical, 70
Variables, deleting, 42–43
Variables, lagged, 46
Variables, measurement, 69, 71
Variables, random, 28–30
Viewing labels, 32

**W**
Weight, robust regression, 264
Whiskers, 112–113
Width of bins, in histograms, 84
Wilcoxon Signed-Ranks test, 194–195, 201
Windows 95
    Preface for Windows 95, xii
    Requirements and limits, 2
    Windows 95 dialog boxes, 10
Windows editor, 27–32
Windows in SQ, 5–13
Windows, arranging for diagnostic plots, 243
Windows, changing, 5–8
Word processor, placing results and graphs in, 14
Working with windows, 9–13

**X**
X-bar + R charts, 168–170
X-bar charts, 164–167

**Y**
Y-hats, regression, 240
Yhats, saving, 248–249

# StataQuest Software License Agreement

Notice to Users: **Read this Notice Carefully**. Do not open the package containing the diskettes until you have read and agreed to this licensing agreement.

## License

StataQuest software is copyright by the Stata Corporation and is protected by United States copyright law and international treaty provisions. This software may be used by more than one person and it may be used on more than one computer, but no more than one copy of this software may be used at the same time. Just as a book may not be read by more than one person at a time without violating the license, this software may not be used by more than one person at a time without violating the license.

Duxbury Press authorizes the purchaser to make one archival copy of the software for the sole purpose of backing-up the software in order to protect the purchaser's investment from loss.

## Limited Warranty

The warranty for the enclosed disks is for ninety (90) days from the original purchase. If, during that time, you find defects in the workmanship or material, the Publisher will replace the defective item. Neither Duxbury Press nor Stata Corporation provide any other warranties, expressed or implied, and shall not be liable for any damages, special, indirect, incidental, consequential, or otherwise.

## No Warranty at all if purchased USED!

StataCorp provides free technical support for the full professional version of Stata. Duxbury Press provides free technical support to StataQuest 4 customers.

| | |
|---|---|
| Voice: | 1-800-423-0563 |
| Fax: | 1-606-647-5045 |
| E-mail: | support@kdc.com |